PHYSICS FOR POETS

PHYSICS FOR POETS

ROBERT H. MARCH

CONTEMPORARY
BOOKS, INC.
CHICAGO

Library of Congress Cataloging in Publication Data

March, Robert H., 1937–
   Physics for poets.

   Reprint. Originally published: 2nd ed. New York :
McGraw-Hill, c1978.
   Bibliography: p.
   Includes index.
   1. Physics.   I. Title.
QC23.M334   1983   530   82-22147
ISBN 0-8092-5532-4 (pbk.)

This book was set in Baskerville by Monotype Composition Company, Inc.
The editors were C. Robert Zappa and Michael Gardner;
the designer was Anne Canevari Green;
the production supervisor was Dominick Petrellese.
the new drawings were done by E. H. Technical Services.
R. R. Donnelley & Sons Company was printer and binder.

Published by Contemporary Books, Inc.
180 North Michigan Avenue, Chicago, Illinois 60601
Manufactured in the United States of America
Library of Congress Catalog Card Number: 82-22147
International Standard Book Number: 0-8092-5532-4

Published simultaneously in Canada by
Beaverbooks, Ltd.
150 Lesmill Road
Don Mills, Ontario M3B 2T5
Canada

This edition is published by arrangement
with McGraw-Hill, Inc.

Contents

Preface

eaders and teachers familiar with the original edition of this book will find the current version much changed, with more than 30 percent new material. The goals of this revision were to make the book less formal, philosophical, and mathematical and more intuitive, historical, and physical. In addition, the treatment of the modern era has been brought up to the present by chapters on general relativity and the quark model. To make room for them, several topics from classical physics have been dropped.

The new approach is due in part to maturation of my own point of view, in part to further experience with teaching the material, and in part to new developments in physics that have produced results sufficiently striking to be of interest to nonspecialists.

One serious defect of the first edition, in my view, was that while it was never intended for a course in which computational skill is a major goal, the text and especially the appendix contained enough mathematics to promote the illusion that it could be so used. The mathematics has been cut back severely in the new edition.

Thus teachers contemplating the use of this book should be warned that they will have to abandon the common (and by no means entirely unjustified) prejudice that problem solving is the soundest test of physical understanding. The course on which this text is based has never functioned on that basis and has moved further from it as it has evolved.

I would like to thank the Stanford Linear Accelerator Center for their hospitality during the period this edition was in preparation. Heartfelt thanks and humble apologies are due my coworkers on the Iron Ball project, for tolerating my preoccupation, irresponsibility, and surliness while the writing was in progress. Special gratitude is due members of the SPEAR Magnetic Detector Group; close proximity to a team of gifted researchers who were transforming physics contributed greatly to the mood of exhiliration in which this book was written.

Though a work like this may be the product of a single pen, it inevitably reflects the thought of many minds. Conversations with students and colleagues, notably Kent Lesandrini in the former category and Charles Goebel and Kenneth Lane in the latter have been invaluable. I have been much influenced by the works of Milič Čapek, Paul Foreman, Max Jammer, and Dennis Sciama. Finally, as any reader familiar with his works will instantly recognize, my debt to Gerald Holton is incalculable.

Robert H. March

PHYSICS
FOR
POETS

Introduction

To the laboratory then I went. What little
right men they were exactly! Magicians
of the microsecond precisely wired
to what they cared to ask no questions of
but such as their computers clicked and hummed.

It was a white-smocked, glass, and lighted Hell.
And there Saint Particle the Septic sat
lost in his horn-rimmed thoughts. A gentlest pose.
But in the frame of one lens as I passed
I saw an ogre's eye leap from his face.

—JOHN CIARDI, FRAGMENT
Saturday Review, April 30, 1966

T HIS BOOK IS devoted to physics for poets. As such it may fittingly be opened with a poem, introduced as Exhibit A, evidence of the need for a book entitled *Physics for Poets.* Not that this book will try to prove that physicists are just like everyone else. Physicists are *not* regular fellows—and neither are poets. Anyone engaged in an activity that makes considerable demands on both the intellect and the emotions is not unlikely to be a little bit odd.

Like many poets, the physicist feels he is looking for "truth." Of course, he defines truth by his own set of rules, and he doesn't think very much about what those rules are (until he gets old, when good physicists often turn into bad philosophers). Thus, he may be just as surprised as the poet to hear that some of those rules have to do with beauty. An idea must be more than right—it must also be pretty if it is to create much excitement in the world of physics. Creativity in any field has an emotional dimension. This may seem surprising, in view of what we are always told about the rules of scientific objectivity. But these rules only concern the way in which an idea gets its final test. The way in which a new idea arises is usually quite the opposite of objective. And if the idea strikes the audience as beautiful, it is likely to be believed even in the absence of confirming evidence and clung to tenaciously until the evidence against it is overwhelming. The creator of an abstract scientific idea has as much of his personality in it as any artist.

In the current era of emphasis on science education, it has become a cliché to call scientific research a great adventure. Well it may be; but the student approaching his first hard science course with this maxim in mind is in for a rude shock. Rarely does much of the sense of adventure manage to come through the hard work, for the subject matter often seems both difficult and dull. The student headed for a scientific career is usually told that he must face years of diligent drill before he can understand anything really profound.

But one wonders how many people would love music if they were required to master a good deal of piano technique before they were allowed to listen to, for example, the Beethoven sonatas. True, a concert pianist probably enjoys the sonatas on some levels denied to others, but a reasonably sensitive person with totally untrained fingers can appreciate their beauty. And the analogy with music may not be as farfetched as it seems. To carry it further, this book will let you listen to a bit of Bach and then see how you make out with Schönberg and Bartók.

Of course, we will have to give up something. What this book gives up is nothing less than a well-rounded view of physics. Much of what physicists traditionally consider important and interesting receives only passing mention or is omitted altogether. Heat, sound, optics, and electromagnetism are the principal victims. Instead, we will concentrate on classical mechanics, relativity, and the quantum theory.

Physics has seen two periods of rapid change. The word *revolution* has been much abused of late, but it is probably appropriate. The first revolution occupied most of the seventeenth century and was so complete that almost nothing preceding it can be recognized as physics at all, in modern terms. The second occupied the first three decades of the current century, and we have clearly not seen the end of it.

It is convenient to regard the first revolution as beginning with Galileo and culminating with Newton (with some injustice to many worthwhile predecessors and contemporaries of these two heroes). It created classical mechanics, probably the most successful scientific theory of all time. For two centuries, this theory swept all before it, one phenomenon after another yielding to explanation in mechanical terms. At the end of the nineteenth century, it seemed on the verge of absorbing optics and electromagnetism and achieving the final unity of all physics. Indeed, to many scientists of that time, it appeared already to have done so, except for a few minor details. But on these last details it ultimately failed—and failed catastrophically.

The triumph of Newton's mechanics had wide repercussions. Leaving aside the legion of (mercifully) forgotten nineteenth-century theologians who came to look upon the Creator as a sort of master clockmaker, we see that a number of intellectual trends developed in response to the success of mechanics. To many of the nineteenth century's philosophers, and even to some of the more influential political thinkers, physics became the model for what an intellectually respectable theory should be. This was probably unfortunate, for Newton's mechanics was in many respects unique. There has never been another theory quite like it, in physics or elsewhere.

Indeed, it has nearly survived the second revolution—for Newton's most important concepts are still part of the language of modern physics. But classical mechanics survives only in an embalmed state, for we now know that it can never again aspire to universality. It rules supreme in a limited domain, but physics has passed it by and has struck out in new directions.

The second revolution has, in fact, compounded the confusion by striking out in *two* new directions: relativity and quantum mechanics. The former was largely the creation of one man, Albert Einstein. The latter grew from the contributions of many thinkers (including Einstein). Relativity is popularly regarded as bizarre and abstruse, but the quantum theory is far more so. Both theories were conceived, at least in part, in much the same spirit—that of critical evaluation of the process by which a physicist actually observes the world in which he lives. Both deal mainly with phenomena that lie outside the realm of ordinary experience. It is partly for this reason that they are so difficult to teach—the phenomena themselves are beyond our day-to-day experience. Both theories contain startling concepts that seem absurd or paradoxical, for they conflict with basic intuitive feelings about space and time, cause and effect.

The realm of the quantum theory is the very small, while relativity deals with the very large or the very fast. Where they come together (in the very small very fast world of elementary particles), they have not gotten along too well. Attempts to combine them still have a bit of the appearance of a forced marriage, and it is not yet clear whether the marriage is doomed or merely slow to settle down. In this sense the revolution is incomplete; one tends to feel that these theories must be reconciled or else supplanted.

Part of the difficulty in teaching these new ideas, especially those of quantum mechanics, comes from the peculiar way in which they have developed. Time and again, over the current century, a remarkable pattern of discovery has repeated itself: a lucky guess based on shaky arguments and absurd ad hoc assumptions gives a formula that turns out to be right, though at first no one can see why on earth it should be. Gradually, physicists come to a more or less satisfying interpretation, at least one which satisfies a physicist. They may still feel uncomfortable, but meanwhile the formula cranks out predictions that turn out to be correct, and it is very hard to argue with that sort of success. An explanation clear enough to satisfy a layman may be a long time in coming. Many of these ideas simply have not yet had time to lose their bizarre quality, and physicists are as reluctant as modern painters to devote much effort to explanations of their work for people who seem naïve, philistine, or just not terribly interested. Like any artist, the creative scientist prefers to feel that his work speaks for itself.

Out of consideration for its intended audience, this book must regrettably work with rudimentary mathematics. Some of the beauty of physics is readily apparent only when it is written in its natural language, which is largely mathematical, and a lot of this beauty is unavoidably lost in translation. To ask a layman to study mathematics merely to appreciate physics is as unreasonable (or as reasonable) as asking him to study Italian merely to properly appreciate Dante. Of course, like Italian, mathematics is beautiful in itself and is likely to be useful for a variety of other purposes.

The worst possible attitude with which to approach the study of physics is one of awe. Like most successful human ventures, physics has prospered largely by sticking strictly to business. There are problems that lend themselves to the physicist's methods, and the solution of them can enrich the human experience, both materially and intellectually. But the scientific method, powerful as it is in its own domain, is neither universal nor magic. Most of what man values and holds significant must remain beyond its scope. If physicists have achieved much, it has been by limiting their ambitions to that which falls within their capabilities.

CHAPTER ONE

Falling Bodies and the Birth of Mechanics

SIMPLICIO: *Your discussion is really admirable; yet I do not find it easy to believe that a bird-shot falls as swiftly as a cannon-ball.*

SALVIATI: *Why not say a grain of sand as rapidly as a grind-stone? But, Simplicio, I trust you will not follow the example of many others who divert the discussion from its main intent. . . .*

—GALILEO, TWO NEW SCIENCES

*I*T IS NOT UNREASONABLE to date "modern" physics from the publication, in 1636, of Galileo's *Two New Sciences.* The title refers to two studies, one on the strength of materials and the other on mechanics, the science of motion. In the case of mechanics, the outstanding achievement was a successful quantitative description of the motion of freely falling bodies. Not only was this description sophisticated and exact, but it also introduced the first quantitative concept for measuring change of state of motion, that of acceleration. Later, Sir Isaac Newton was to put this to good use.

In popular history, Galileo is often portrayed as a lone seer, the only objective observer in the midst of a pack of fools willing to trust the authority of ancient philosophers rather than the evidence of their own eyes. But were you to choose at random 10 reasonably portable objects from the room in which you are now sitting and cast them out the highest available window, the chances are that few of them would fall in a manner that much resembles the simple motion described by Galileo.

His description of free fall was an idealization, an insight that neglected most of the complexity of the fall of real objects. His adversaries held to a description inherited from Aristotle. While this description had serious flaws, for many examples of fall it came closer to describing the phenomenon than Galileo's did. But Aristotle's insight was no more than a lucky empirical guess, a scientific dead end, while Galileo had taken a giant step toward a deep understanding of motion.

THE FIRST MODERN PHYSICIST

Galileo lived from 1564, the year of Shakespeare's birth and Michelangelo's death, to 1642, the year Newton was born. He was born in Pisa to a somewhat notable but impecunious Tuscan family. His father, a musician and amateur scholar who wrote one of the first modern treatises on harmony, hoped his clever son might recoup the family's fortunes by means of a medical career. But at the university, Galileo became fascinated with mathematics, then as now a far less lucrative profession.

Though he is a hero to all modern scientists, even his greatest admirers must admit that Galileo was often boorish, pugnacious, and petty. He was occasionally unscrupulous in seeking his own advancement, on several occasions claiming credit for the work of others. Tradition pictures him as a glutton and womanizer who lived with a bravura that charmed his friends but outraged his enemies. He had a remarkable gift for the written word and could not resist the temptation to sprinkle his works with elegant insults to his opponents. In short, he was very much the late Renaissance man.

Like most present-day scientists, Galileo had his most productive years, in terms of originality of thought, when he was comparatively young. These coincided with his term as professor of mathematics at the University of Padua, a post he held from the age of 27 to 46. The liberal atmosphere of the Republic of Venice, of which Padua was the second city, was hospitable to thinkers of Galileo's style. His university and its neighbor in Bologna, the oldest two in Italy, were practically the only schools in the world where *natural philosophy,* as the physical sciences were then called, was taken seriously. He had been forced to flee his native Pisa, where orthodoxy was more prized and insults to respected scholars were not tolerated. But in his middle years, he was lured back to Florence by Cosimo di Medici, a former student of his who then ruled Tuscany. The bait was an unprecedented salary and freedom from formal teaching duties; thus Galileo's academic career was every bit as modern as his scientific method.

Galileo's brashness proved a severe liability in the subtle politics of the Florentine court. Within 6 years his powerful enemies made good

Portrait of Galileo by Sustermans. (The Granger Collection.)

use of the hysterical response of the Roman church to the Protestant Reformation, securing an edict denouncing his work supporting Nicholaus Copernicus' heliocentric universe. But with a man of Galileo's towering reputation, the Church was reluctant to deal in a cavalier manner. The edict was little more than a nuisance, forcing him to treat the copernican theory as a hypothesis to which he did not necessarily subscribe. With the accession to the papal throne of his longtime friend Cardinal Barberini (Urban VIII) in 1623, Galileo was emboldened. Here might be a chance to cap his career by forcing Church acquiescence to, if not actually adoption of, the heliocentric universe. Accordingly he published, in 1632, his *Dialogue on the Great World Systems*. The work was widely acclaimed almost from the day of publication, and the tone was Galileo at his sardonic best. The book was cast as a platonic dialogue, in itself an insult to the scholars of his time, who revered Aristotle and imitated his careful analytic style. Moreover, their best arguments were assigned to a character named Simplicio, a country bumpkin whom Galileo handled rather roughly.

Urban was in trouble himself over his personal ambitions and the failure of his military adventures. Setbacks in the Thirty Years' War had forced him to court alliances with Protestant princes. He was thus in no position to come to the aid of an old friend of questionable orthodoxy. Furthermore, Galileo's enemies whispered that Simplicio could be taken as a caricature of Urban himself. The Pope stepped aside and let the Inquisition do its work. Under threat of torture, Galileo recanted his heretical views. The last decade of his life was spent under the watchful eye of the Church, though his work continued to appear abroad in editions he was obliged to denounce. This period saw the publication of his greatest work, the *Two New Sciences*.

A present-day physicist reading the work of Galileo can hardly fail to recognize in him a colleague, one whose style of thought and argument are fully contemporary. Were a time machine to deposit Galileo in a twentieth-century university physics laboratory, it is easy to imagine him buckling down to work on the most interesting problems. More than likely, he would find university deans, foundation officials, and most of his fellow physicists as repugnant as he found their Florentine counterparts. His attitude toward the value of experiment and his reluctance to bring up questions more general than demanded by the data on hand are exactly those attitudes drilled into young researchers in the course of study for a Ph.D.

The era in which Galileo worked was quite hospitable to the emerging natural sciences. The late Renaissance had elevated the lay scholar to a high station in public life. The recent invention of the printing press had opened the way to wide, rapid dissemination of new ideas. This destroyed the monopoly of learning of the academies and monasteries, where close personal contact and isolation from lay support helped to keep heretical scholars in line. A worldly humanism, with an infinite faith in the power of man's reason, and a pagan love of life constituted the spirit of the times. Galileo had the effrontery to publish in the vernacular Italian rather than the Latin of a respectable scholar. His opponents could hardly overlook the implied insult; Galileo had left the narrow confines of established scholarly debate and taken his case to a public he regarded as more open-minded than the established authorities of the academies. Even today, a scholar whose ideas meet with a poor reception in academic circles will be in considerable trouble if he decides to promote his views in, for example, the science columns of the daily press.

But, repugnant as it may have been to the academic establishment, Galileo's style of work was well established by this time. A group of scholars with comparable methods of work had formed several nascent scientific organizations. Galileo was a member of one of these, the Academia dei Lincei, which had a kinship both with a learned society and a secret fraternity. Its members met to dine and debate, fostered scientific correspondence, and aided members and their protégés in securing publication of their works. Thinkers of this stripe had many places to turn to for financial help, partly because of the general intellectual atmosphere and partly because they were the first to bring the analytical tools of scholarship to bear on the practical problems of the artisan, thereby creating the modern ties between science and technology.

GALILEO TAKES ON ARTISTOTLE

The central intellectual event of the early Renaissance had been the rediscovery of classical Greek philosophy, particularly the works of Aristotle. The Greek intellectual heritage had been preserved through the Middle Ages by the Arab civilization, which until at least the thirteenth century had every right to look down on Europe as a barbarous backwater. In the early stages of Europe's recovery, such scholars as St. Thomas Aquinas molded ideas taken from Greek thought, notably the work of Aristotle, into an all-encompassing world view known as *scholasticism*. The scholastics were most concerned with moral philosophy and theology and added little to the Greek achievements in physics. These had centered on static problems; the Greeks had made only a modest start in dealing with motion. Thus the scholastic philosophy of nature emphasized the static order of the universe, with every object in its proper place, revealing perforce the wisdom of the Creator. Motion was a temporary and possibly unnatural state, not totally unworthy of study but certainly of secondary importance. A stone fell because it sought its natural place, on the ground; flames lept up to seek reunion with the divine fire of the stars. To inquire into the messy quantitative details of such processes when their larger cosmic role was already well understood was regarded as a sterile exercise. In the three centuries between Aquinas and Galileo, scholasticism had frozen into a dogma of almost scriptural rigidity. Its proponents felt smugly that it encompassed nearly all that was worth knowing, and respectable scholars spent their lives myopically working through the words of Aristotle.

It would be unfair, however, to accuse either Aristotle or his medieval followers of ignoring experimental evidence. Aristotle was a careful, systematic observer who believed that order was there in nature, waiting to be uncovered through observation, comparison, and classification. The great descriptive sciences of the nineteenth century, such as botany and zoology, were entirely aristotelian in their methodology.

Galileo's methods, however, were more subtle. As revealed through his notebooks, he was certainly a remarkably skillful experimenter. But unlike his present-day counterparts, he did not accept experiment as the final arbiter of truth. Galileo considered himself a follower of Plato, who felt that order resided not in nature but in the human mind. Only pure reason could rise above the fallibility of the senses to uncover a higher, more perfect reality than any to be found in nature itself.

But Galileo's platonism had an empirical twist. From personal experience, he had learned that pure reason, unaided by the senses, can sometimes be led astray. Careful observations were a good means of rooting out error and placing reason back on the right track. Experiments were also useful for convincing doubters who could not follow his arguments or were too stubborn to accept them. But he never expected nature to mirror perfectly the higher truths of science or to reveal them to the passive observer. Under carefully selected conditions, the deviations from the ideal state could be kept small enough to be ignored safely.

It is this curious tension between empiricism and idealism that makes modern experimental science possible. It frees scientists to follow their hunches toward deeper truths without becoming mired down in a myopic concern for detail.

Aristotle had not completely ignored motion. In particular, he had formulated a quantitative description of the motion of falling objects. He asserted that if one compares bodies falling in the same medium, one will find that they fall with speeds proportional to their weights.

From the point of view of a physicist, this was a very good hypothesis, not because it was right, but because it could be either very right or very wrong. A quantitative statement of the sort Aristotle had made is valuable because it really commits a theory to a severe test. If I were merely to predict that the sun will rise tomorrow, the statement would hardly cause much excitement. But if I predicted that it would rise tomorrow exactly 1 minute (min) and 32 seconds (s) later than today, I would leave myself open to being proved wrong by a trivial check of a watch. It is a cardinal rule of the scientific method that a hypothesis is useful only if it can in principle be proved wrong. Speculations that are not testable are regarded as "unscientific."

Demolishing Aristotle's falling-body law took little effort on Galileo's part, because some of its predictions are so wrong that they can be easily refuted. Indeed, its falsity had been recognized at least as early as the sixth century by the philosopher John Philoponus. Thus, by adhering to every word of Aristotle as second only to the scriptures in authority, the scholastics left themselves open to demolition by appeal to everyday experience. In the following excerpt from *Two New Sciences,* Salviati, the author's spokesman, does exactly this with the hapless Simplicio and a third interlocutor, Sagredo, a reasonably intelligent practical-minded humanist much like those Galileo sought as an audience when he wrote in Italian.

> SALVIATI: I greatly doubt that Aristotle ever tested by experiment whether it be true that two stones, one weighing ten times as much as the other, if allowed to fall, at the same instant, from a height of say, 100 cubits, would so differ in speed that when the heavier had reached the ground the other would not have fallen more than 10 cubits.
>
> SIMPLICIO: His language would seem to indicate that he had performed the experiment, because he says: "We see the heavier": now the word *see* shows that he had made the experiment.
>
> SAGREDO: But I, Simplicio, who have made the test can assure you that a cannon ball weighing one or two hundred pounds, or more, will not reach the ground by as much as a span ahead of a musket ball weighing only half a pound . . . .

The excerpt is an excellent example of Galileo's forensic style. Not content to merely demolish the Aristotelian theory, the author cannot resist the temptation to get in a dig at the methods of the classical scholar by holding up to ridicule Simplicio's excessive concern for the exact meaning of every word in Aristotle's work. Unhappily, this particular form of scholarly nitpicking is not yet extinct, as any college student can testify.

Aristotle also stated that a falling body instantly acquires its speed of fall; an equally simple argument serves to dispose of this prediction:

> SALVIATI: But tell me, gentlemen, is it not true that if a block be allowed to fall on a stake from a height of four cubits and drives it into the earth, say, four finger-breadths, that coming from a height of two cubits it will drive the stake a much less distance, and from the height of one cubit a still less

distance; and finally if the block be lifted only a finger-breadth how much more will it accomplish than if merely laid on the stake without percussion . . . and since the effect of the blow depends on the velocity of the striking body, can any one doubt the motion is more than small whenever the effect is imperceptible?

Of course, the argument does depend on the unproved (but reasonable) assumption that the speed of the falling block is what determines its effectiveness as a pile driver, but how explain the phenomenon in Aristotle's terms, where the distance fallen should have no effect whatsoever?

However, tearing down is always easier than building up, and there had been many critics of Aristotle. Galileo earned his present place in scientific esteem by offering his own description of the motion of falling bodies: "In a medium totally devoid of resistance all bodies will fall at the same speed . . . [and] . . . during equal intervals of time [a falling body] receives equal increments of velocity. . . ."

The words "totally devoid of resistance" were crucial in this description; they represent the abstraction from nature that led to Galileo's success, and they provided a means of countering Simplicio's observation (see the heading of this chapter) that heavier bodies do indeed fall somewhat faster. Those words were a daring innovation, because "totally devoid of resistance" implies a *vacuum*. Not only was it impossible in practice to achieve a vacuum in Galileo's time, but the prevailing scientific thought regarded a vacuum as a most unnatural state: "nature abhors a vacuum." Furthermore, some ancient thinkers about motion regarded the medium in which a body moved as essential to supply the motive force; Simplicio at one point is depicted as expressing doubt that motion can even take place in a vacuum.

Nonetheless, Galileo could not cavalierly dismiss Simplicio's objections. Feathers indeed fall more slowly than cannonballs. Galileo was sure the effect was due to retardation by the medium, but it was beyond his powers to prove this conclusively. He was forced to offer arguments that merely made it seem plausible:

Have you not observed that two bodies which fall in water, one with a speed a hundred times as great as that of the other, will fall in air with speeds so nearly equal that one will not surpass the other by as much as a hundredth part? Thus, for example, an egg made of marble will descend in water one hundred times more rapidly than a hen's egg, while in air falling from a height of twenty cubits, the one will fall short of the other by less than four finger-breadths.

In short, if the deviations from his law are far worse in dense media than thin ones, is it not reasonable to suppose they disappear if the medium is absent altogether?

Of course, it is always preferable to be able to account in a precise fashion for deviations from a proposed scientific statement, but failing this, an argument like that above is usually convincing enough. Science is far less absolute in its requirement of agreement with experiment than is commonly supposed, and nearly all contemporary scientific papers will contain at some point the sort of reasoning used by Galileo to deal with this case.

As a final touch, Galileo insisted, in a quite sophisticated fashion, that the question of the possibility or impossibility of a vacuum was quite irrelevant to the validity of his law. This is a very modern point of view, yet at the same time very ancient, dating from Plato and Socrates. It is possible to understand nature in terms of approximation to an ideal state even if that state cannot possibly exist in nature.

But the acid test of Galileo's law lay in his assertion that speed increases with time of fall. To demonstrate this he had to invent some of the modern mathematical language to deal with motion. This is the real starting point of modern physics, so we must pause to develop the necessary concepts.

THE MATHEMATICAL LANGUAGE OF MOTION

Speed and *velocity* are terms familiar to anyone raised in the twentieth century. Actually, the two terms have distinct meanings in exact scientific parlance, which will be explained further in Chap. 3, but even physicists slip and use them interchangeably. Speed may be defined as follows:

$$\text{Average speed} = \frac{\text{distance moved}}{\text{time elapsed}}$$

The qualification "average" is necessary in recognition that the speed may not have been the same in all parts of the time interval. Expressed more concisely in conventional symbols, our definition becomes

$$\bar{v} = \frac{\Delta s}{\Delta t} \tag{1-1}$$

Here the Greek capital delta (Δ) has been used to denote change or interval, reminding us that the measurement compares two different positions at two different times. Of course, we could have used one-letter symbols for Δs and Δt, but this common device serves to jog the memory and more fully convey the significance of the formula. Similarly, the bar over the v is one of several conventional ways of saying "average value of" whatever symbol is found beneath. The use of s as a symbol for distance is conventional, for some obscure reason. Part of the role of mathematics in science is as an extension to the language, and notation is chosen for reasons of clarity.

If you were to use Eq. (1-1) to compute the speed of your car in, say, $\frac{1}{2}$ hour (h) of city driving, taking odometer readings and the time on your watch, you might obtain 17 miles per hour (mi/h). But this could conceal 5 min at a dead halt in a traffic jam and a few daring seconds at 70 mi/h. A lot of detail of the motion is obviously left out. To make the description more complete, you might further subdivide the trip into 1-min intervals, calculating the average speed for each interval. When must this process stop?

The physicist's answer is pragmatic: When I have enough detail to answer the question at hand. As a rough guide to the harsh realities of modern city driving, your original measurement was probably pretty good. But suppose you want the kind of figures that enable you to tell how well your car performs? A physicist would reason as follows. A car can change its speed by perhaps 10 mi/h in 1 s when accelerating or braking hard. Thus, if I choose intervals of $\frac{1}{10}$ s, the speed at the beginning of the interval will differ from that at the end by at most 1 mi/h. In 1 millisecond (ms = $\frac{1}{1000}$ s) I can get a result good to within $\frac{1}{100}$ mi/h, and so on.

Figure 1-1 serves to illustrate the problem graphically. If an object is moving at constant speed, the relation between distance traveled and time elapsed is a straight line. The steeper this line, the greater the speed. The steepness of a graph is called its *slope*. When speed is changing, the slope is also changing. The distance-time graph of a motion at changing speed is thus a curved line.

When the speed is not constant, the measurement of an average speed leaves out much detail of the graph. In Fig. 1-2 we see that

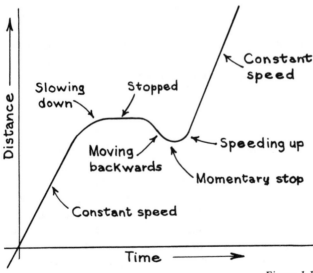

Figure 1-1

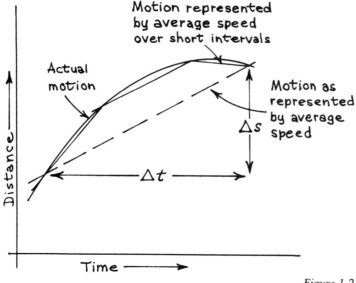

Figure 1-2

measurements over smaller intervals more nearly represent the actual motion.

A perfect description, of course, would require infinitely small intervals. This is in effect what the calculus is all about, but we need not be that formal. The essential point to grasp is that no matter how accurate a description you want, it can be obtained simply by choosing a small enough time interval. This assures you that it is meaningful to talk about the speed at a particular instant. An automobile speedometer, in fact, measures this speed directly. In this case, we can drop the qualifying adjective *average*.

The next step is to find a way to indicate how fast the speed itself changes, as when an automobile firm claims that its product will do "zero to sixty in six seconds." We can express something similar in the same way we defined velocity, getting a new quantity called the average acceleration

$$\bar{a} = \frac{\Delta v}{\Delta t}$$

By this definition, the acceleration claimed in the car ad is 10 mi/h per second. A physicist would shudder at using hours as the time unit for speed and seconds for acceleration and would prefer some unmixed unit such as meters per second per second, which is contracted to meters per second squared (m/s²). As in the measurement of velocity, Δv is calculated by subtracting the speed at the beginning of the time interval from the speed at the end.

Acceleration is a difficult concept to grasp by virtue of being two steps of abstraction removed from the direct experience of seeing something in a particular place at a particular time. To use an economic analogy, the price of eggs at the supermarket is a concrete fact. The statement that the rate of inflation is 10 percent per year is an abstraction, in fact the same one involved in measuring speed. The doleful news that things are getting worse because 2 months ago the annual inflation rate was only 8 percent is the next level of abstraction, exactly analogous with acceleration. If we divide the change in inflation rate, which is $10 - 8 = 2$ percent, by the time interval between samplings, which is $\frac{1}{6}$ year, we have the "acceleration" of prices, 12 percent per year squared, and the analogy is complete.

The concept of acceleration handles equally well processes of speeding up and slowing down. In the latter case the change in velocity (and therefore the acceleration) is negative. So we need no separate concept of *deceleration*.

It is important to keep the distinction between speed and acceleration firmly in mind, as they can be quite independent of one another and can even be opposite in sign. When a car is moving forward and speeding up, both speed and acceleration are positive. If it is moving forward and slowing down, speed is positive but acceleration is negative. For a car that is speeding up in reverse, both acceleration and speed are negative. But if a car in reverse is slowing down, its speed is negative but its acceleration is positive, because its speed is becoming less negative, which is a change in the positive direction.

If that last sentence seems to make sense, you have a very good head for physics. Referring back to Fig. 1-1, it is clear that when acceleration is present, the line describing the motion must be curved. If the object is speeding up, the graph curves upward. If it is slowing down, it curves down.

THE CRUCIAL TEST

Galileo used the concepts developed above to derive an interesting result which could be compared with experiment to test his theory. He obtained the relation between distance traversed and time elapsed for a uniformly accelerated body starting from rest. His law of falling-body motion is nothing more, in our new language, than the statement that "a is a universal constant for falling bodies." That is to say, if we measure the acceleration of a falling body, we will find it the same at all times during its fall, and the same for all bodies.

When a body starts from rest with a constant, its average speedup to any given moment must be one-half its present speed at that moment. This can be seen by means of a little arithmetic on the numbers in the second column of Table 1-1. The numbers in the table represent a constant acceleration (note the equal speed changes in equal time intervals) of 10 m/s², which happens to be very nearly that for a freely falling body. This result may be expressed as

$$\bar{v} = \frac{v}{2} = \frac{at}{2}$$

which reads: "To find the average speed, multiply the acceleration by the time to get the final speed and divide by 2," because the body was speeding up and thus was going faster at the end than at all times earlier. To get the distance traveled in a time interval, we multiply the average speed by the time:

$$s = \bar{v}t = \frac{at}{2}t = \frac{at^2}{2}$$

In interpreting this formula, we should always remember that the factor ½ comes from the fact that the average speed is half the final speed and the time is squared because it enters in twice; it allows the body to speed up and also allows it to move farther at whatever speed it is going.

　　To demonstrate his proposition that falling bodies acquire equal velocity increments in equal times (which is tantamount to showing that the distance traveled varies as the square of the time), Galileo faced serious experimental difficulties. With the best time-measuring instruments of his era, he could scarcely measure intervals to a fraction of a second. Yet a heavy object dropped from a tower 150 ft high will reach the ground in only 3 s!

TABLE 1-1

Time, s	Speed, m/s	Distance fallen, m
0	0	0
1	10	5
2	20	20
3	30	45
4	40	80
5	50	125
6	60	180
	Average speed = 30 m/s	

To solve this problem, Galileo chose to study the roll of a ball down an inclined plane. By ingenious arguments he asserted that this would "dilute" the motion of a falling body (i.e., reduce the acceleration) without fundamentally altering its character. This assertion had to be taken somewhat on faith, because Galileo did not have a complete theory to show exactly what effect the inclined plane would have. But a new theory in physics is rarely complete as first presented; there are often large logical holes that must be filled in later.

Using a smooth board with a small tilt and a groove to guide the ball, Galileo was able to produce indoors a motion that took about 10 s to complete. His timer was crude but adequate for this experiment: a vessel of water with a hole in the bottom that he closed by means of his finger. When the finger was removed, water flowed into a cup. Afterward, the cup was carefully weighed. The amount of water in the cup was a measure of the time. The results agreed with Galileo's prediction. To argue that this result had any bearing on the problem of free fall was of course a bit of a logical leap. But ideal experimental conditions are hard to come by, and indirect tests, supported by arguments that are plausible but not completely rigorous, play an important role in the development of a young science.

WAS ARISTOTLE SO WRONG AFTER ALL?

The scholastics might have fared better in arguments with Galileo had they had a spokesman capable of using his style of argument. Let us explore the question of falling body motion as Galileo's adversary for a while. If we were actually to observe the fall of a body from a great height, measure its speed at all times, and present the results in a graph, we would obtain the curve shown in Fig. 1-3. The reason for this curious behavior is very simple. As a body speeds up, the resistance of the air to its motion increases. Eventually, a speed is reached where the force resulting from the rush of air matches that pulling the object down and no further acceleration takes place. This speed is called the *terminal velocity* of the object. Interestingly enough, if we compare bodies of the same size and shape, their terminal velocities are very nearly proportional to their weights, as in Aristotle's falling-body law. A heavy steel ball falling from an airplane might require thousands of feet to achieve terminal velocity; a human body acquires it in a few hundred feet—the secret of sky diving, which is a long fall at terminal velocity, followed by the opening of a parachute to lower the terminal velocity to a safe value for landing.

Now it is far easier to study the motion of a body with a low terminal velocity, for example, a light object falling in a dense medium, such as a golf ball in water. This kind of object is what Aristotle spent

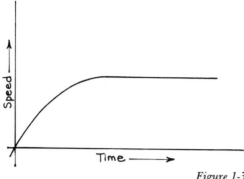

Figure 1-3

most of his time discussing. There is no a priori reason for assuming this approach to slowing down falling-body motion to a reasonable speed is any less legitimate than Galileo's choice of an inclined plane. An object can acquire terminal velocity in a fraction of a second in a dense medium, and a graph of its speed versus time might look like the curve in Fig. 1-4.

The graphs look even more striking when we consider that actual measurements must of necessity by of *distance* and time. A real experiment, if made with instruments available in Galileo's time and with its results presented in graphical form, as is customary in scientific journals, might look something like Fig. 1-5.

From these data a reasonable person might conclude that Aristotle was closer to the truth than Galileo. All we have to do is admit that the process of acquiring speed is not quite instantaneous, and Aristotle is out

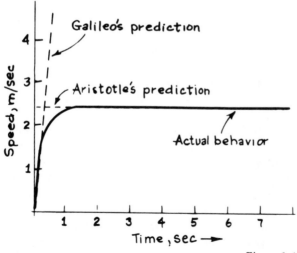

Figure 1-4

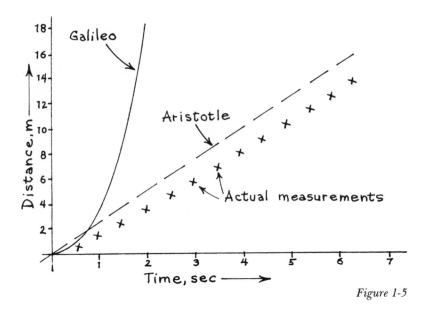

Figure 1-5

of the woods, with only a small modification of the original hypothesis. Add to this the notion from classical philosophy that the medium is in fact the source of motive power for the descent and that the vacuum which Galileo regards as *his* ideal situation is a most unnatural state, and we might reasonably conclude that Aristotle must be dealing with the more fundamental situation, while Galileo has been led astray by his excessive concern with a short-time phenomenon which goes away if we watch long enough. "Surely," we might tell Galileo, "you have allowed yourself to be misled by the fact that heavy objects falling in air for some reason take a long time to reach their natural speed. Besides, the only good data you have come from the inclined plane, which may not be relevant to the problem of free fall."

But, it turns out, this quite reasonable point of view is wrong; not that it is a worse approximation to nature, but the regularity observed by Aristotle proves to have a far less fundamental significance than that observed by Galileo, viewed in the context of the subsequent development of the science of mechanics.

Indeed, Galileo described experiments with falling bodies in fluids and realized there was some value in Aristotle's work, though he was vain enough to attack him on a minor point—that Aristotle did not take into account the difference in weight between an object submerged and the same object in air, an effect discovered after Aristotle's time by Archimedes. With this correction, Aristotle's prediction would have worked out better, but Galileo, with his debater's instinct, insisted on holding Aristotle's latter-day followers to the original version, and they were too hidebound to strengthen their case by updating the theory.

Nonetheless, Galileo reasoned that the motion of falling bodies in fluids was sensitive to too many minor factors, such as the shape and the size of an object as well as its weight, to represent an important fundamental rule of nature. He found his *own* regularity, a universal speed and a uniform acceleration, far more appealing as a law of nature, and subsequent developments bore out his insight. But it represented something more than mere observation and shows that the vaunted objectivity of science is not as naïve as it appears on the surface.

The message to be learned from this exercise in devil's advocacy is that there is nothing automatic about scientific progress. Suppose this particular problem were currently on the research frontier. A well-trained modern Ph.D. physicist might well receive a large research grant to study falling bodies, which he could use to acquire large quantities of data on falling objects in all possible combinations of shape, size, weight, medium, etc. He and his students would strive to refine their measurements, producing a deluge of papers for the scholarly journals. It is not inconceivable that, faced with the necessity for explaining all these data to a reasonable degree of accuracy without a complete theory, they would move toward a point of view like that of Aristotle and overlook Galileo's insight altogether. Science is more than a mere attempt to describe nature as accurately as possible. Frequently the real message is well hidden, and a law that gives a poor approximation to nature has more significance than one which works fairly well but is poisoned at the root.

CHAPTER TWO

Toward a Science of Mechanics

If I have seen farther than others, it has been by standing on the shoulders of giants.

—SIR ISAAC NEWTON

*M*ERE DESCRIPTION, no matter how precise, is only a primitive first step in the development of a science. Galileo understood this thoroughly when he dismissed his own work on falling bodies as merely "some superficial observations." Describing one form of motion, he realized, was of little value in dealing with the problem of motion in general. There was a clear need for general principles applicable to many or all forms of motion.

A few such principles had been proposed by Galileo's time, and he had the insight to choose the most significant, refine them, and finally demonstrate their power by solving the difficult problem of projectile motion. Then, a few decades later, the French philosopher René Descartes formulated a truly original law, the value of which was demonstrated in his treatment of collisions between objects. These two achievements are the topic of this chapter. They brought mechanics from the descriptive stage to the *phenomenological* phase of its development. When a science reaches this stage, general principles are used to find connections between a limited number of phenomena but these principles do not yet constitute a complete theory. That latter step had to await the work of Newton, the hero of the next chapter.

Galileo's claims for his ideas on mechanics were modest, but they reveal a conceit that underlies the conservatism of the cautious empirical methods employed by modern scientists: "There have been opened up to this vast and most excellent science, of which my work is only the beginning, ways and means by which other minds more acute than mine will explore its most remote corners."

PROJECTILES FOLLOW PARABOLAS

Galileo used two simple principles in his analysis of projectile motion:

> THE PRINCIPLE OF INERTIA: *A body moving on a level surface will continue in the same direction at constant speed unless disturbed.*

> THE PRINCIPLE OF SUPERPOSITION: *If a body is subjected to two separate influences, each producing a characteristic type of motion, it responds to each without modifying its response to the other.*

The principle of inertia, like the description of falling-body motion, was a choice between two extreme ways of idealizing a complex phenomenon. The motions we observe in the real world all have some tendency to continue after the agent causing the motion is removed, but the motion persists only for a limited time. To cite two extreme examples, consider a stone dragged across rough ground or a chunk of ice sliding across a frozen lake. Most ancient thinkers generalized from the former case and dismissed the persistence of motion as a temporary condition. By Galileo's time, however, the men who founded modern physics preferred the opposite generalization: motion has a natural tendency to persist unless the roughness of the surface interferes.

Each point of view had some appeal to the intuition, and there was no obvious basis for a choice between them. Once again, the most significant test in the final reckoning was not which idea more nearly described the motions ordinarily observed in nature but which ultimately led to a deeper understanding of nature. Galileo's approach led straight to the triumphs of Newton, whereas the ancient picture had nowhere to go.

The acceptance of the principle of inertia completely changed the direction of speculation about motion. It brought a recognition that there is a certain kinship between an object at rest and an object in motion in a straight line at constant speed. This kinship is intuitively obvious to us today, familiar with smoothly moving conveyances like airplanes and ocean liners; it is difficult to detect any signs of motion without looking out the window. But Galileo and his contemporaries had never had such an experience. Thus, it was quite a feat to come to the realization that it was not motion itself but deviation from simple constant motion for which a cause had to be sought. The essence of the experimental method is that when you finally get around to asking nature the right question, she will give you a simple answer; Galileo had indeed stumbled across the right question, even if the final answer was to prove a bit beyond his grasp.

The principle of superposition leads directly to a simple but surprising result. It tells us that if a gun is fired horizontally and, at the same instant, a bullet is dropped from the height of the muzzle, both bullets will hit the ground at the same time. In the absence of air resistance, the rapid horizontal motion has no effect on the vertical motion. The winging bullet falls at exactly the same rate as the one dropped from rest, and they remain always at the same height until they reach the ground, as illustrated in Fig. 2-1.

The more common case, where the projectile starts off moving upward, is only slightly more complicated. The horizontal motion is at constant speed, while the vertical motion is that of an object thrown straight upward, which rises and then returns to the ground, as shown in Fig. 2-2.

To describe the vertical motion, let us backtrack a bit and complete Galileo's picture by describing the motion of an object thrown upward. Such a body *diminishes* in speed in exactly the same fashion that one traveling downward increases its speed. In the upward portion of the motion, the effect of the acceleration is to reduce the velocity; in equal time intervals, the body *loses* rather than gains equal increments of speed. In any part of its upward travel, it loses exactly as much speed as it would gain in the corresponding downward portion. The ascent looks exactly like the descent would if it were photographed by a motion-picture camera and shown in reverse. The body rises and loses speed until it (instantaneously) comes to rest at the apex of its flight. In a time equal to that required for its rise, it falls back, hitting the ground with the same speed it had when it left the ground.

To analyze this motion mathematically, we must obtain the vertical and horizontal parts of the velocity separately at the start of the projectile's flight, which is a problem in trigonometry. The mathematical details are explained in the appendix. But the idea behind the analysis is simple. The vertical part of the motion determines how long the body

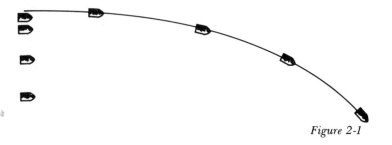

Figure 2-1

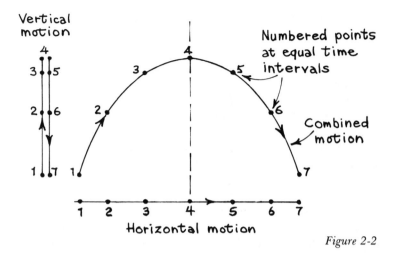

Figure 2-2

will remain in flight. Once the time of flight is known, it is multiplied by the horizontal part of the velocity to find how far away the projectile will land.

The time of flight is obtained by dividing the vertical velocity by the acceleration due to gravity. This tells how long it takes for the body to lose its initial upward thrust and reach the top of its trajectory. The descending portion takes the same time, so we multiply the result by 2 to get the total flight time.

Galileo used this analysis to construct a simple "artillery table," giving the range of a projectile in terms of the speed of the projectile and the angle at which the gun is pointed. The rulers of his time had as lively an interest in military technology as those of our day, but Galileo's table hardly qualifies as a practical example. Since his analysis ignores air resistance, the table is absolutely useless for any practical purpose. Air resistance could shorten the range of a cannonball of Galileo's time by hundreds of yards. Besides, to make practical use of the table, the velocity of the projectile when it left the gun would have to be measured, which was out of the question with the instruments then available.

What was *most* significant to Galileo's contemporaries was that he proved the path of a projectile was a parabola, which we can easily demonstrate. Looking at the trajectory from the apex down, the vertical distance covered is, of course, proportional to the square of the time, since it is the motion of a body falling from rest, which we discussed in Chap. 1. Meanwhile, the horizontal motion is proportional to the time. Thus, the vertical distance fallen is proportional to the square of the horizontal distance covered. Since ancient times it had been known that this relation between vertical and horizontal distances from an apex is the unique property of a parabola. By the motion-picture-in-reverse analogy, the rising portion of the trajectory looks the same. Even though no precise experimental test was then possible, it struck the geometri-cally minded philosophers of Galileo's time as so reasonable that it lent credence to his methods. After all, the parabola is one of the simplest curves known to mathematicians. Any reasonable observer can see that a projectile follows some curve, and a few had even guessed that it might possibly be a parabola. But until Galileo, nobody had been able to attack the problem from a theoretical point of view with anything like a convincing argument. Though the cardinal rule of modern physics, enunciated by Galileo himself, is that arguments are to be settled only by recourse to experiment, even when no experimental test is possible, a seemingly reasonable and pleasingly simple result will attract a lot of support to a theory, for everybody wants to see it come true.

Physicists commonly assume that nature is simple; Newton even later elucidated this credo in as many words, and nearly all physicists take it as an unstated article of faith. But it is, in a sense, a self-confirming hypothesis. If a problem turns out not to have a simple solution, a physicist will come to regard it as not fundamental. When pressed to explain what he means by a "fundamental" law, a physicist is ultimately forced to invoke the essentially aesthetic criterion of *simplicity* or *elegance*. Many physicists have been known to express surprise that elegant laws of nature do exist; it seems nature has been almost too helpful. But physics, at least on its most fundamental level, is a science which has no obligations to the phenomena it studies. If they prove too complex, they are usually dismissed as insufficiently fundamental and forgotten, unless a later generation finds them of practical value and takes the trouble to work through all the detail. Thus, few physicists are even aware of the true path of a projectile, subject to air resistance. It is a complicated curve mathematically difficult to describe, and thereby less "fundamental" than Galileo's parabola, even though the latter is a useless approximation to nature for nearly all practical purposes. Indeed, the details of projectile motion in the air were not really worked out until our own century.

The history of physics contains many such examples of whole fields of study in which the original problem that served as the starting point was never fully solved but was abandoned after it had served its purpose.

DESCARTES, HUYGENS, AND MOMENTUM CONSERVATION

The next major contribution to the development of mechanics came from the French philosopher René Descartes, born one generation after Galileo. While Galileo had made a good start at building mechanics from the bottom up, Descartes tried to work from the top down. His goal was to construct a general philosophy, as a replacement for that of the scholastics, by means of his own meditations and analytical methods that placed great emphases on the discovery and use of *first principles*. Having only the piecemeal education of a young man of the lower gentry trained for a military career, he disdained scholarship and erudition, placing little stock in the work of others. For example, there is no indication that he was even acquainted with the works of Galileo.

In terms of goals and methods, he was in many respects more nearly akin to the ancient philosophers than Galileo, whose style was entirely modern. To this day, French academic training encourages the use of the cartesian style of argument, in which a basic principle is isolated and then followed by impeccable deduction to conclusions of truly astonishing and often infuriating scope, a practice that does not always sit well with more empirically minded scientists in other lands.

Indeed, much of the credit for Descartes' achievements in physics should go to Christian Huygens, the son of a Dutch diplomat, at whose home Descartes was a frequent guest. Though he ultimately rejected Descartes' philosophical system, Huygens salvaged from it those parts that were most useful for physics and corrected some of its more glaring scientific errors.

The achievements of Descartes' mechanics fell far short of his projected goals, but he left two indelible marks on the history of physics. First, he directed attention to the problem of the interaction of two moving objects, an emphasis Newton was to turn to great advantage. Second, in the process of studying this problem, Descartes demonstrated the power of a remarkable format for constructing a law of nature, the *conservation law*.

A conservation law might be called the scientific equivalent of the French aphorism, *plus ça change, plus c'est la même chose* (the more things change, the more they remain the same). Applied to a complex process in which things are constantly changing, a conservation law is an assertion that some simple quantity remains the same. The true power of this kind of law was not fully realized until well after Newton's time. Present-day physicists are so accustomed to thinking in terms of conservation laws that many attempts to formulate new basic laws of physics are phrased in this form.

A conservation law rarely provides a complete description of a process, for in spirit it implies that the details need not be considered—they will work themselves out. Herein lies its power, for it is exempt at the outset from the necessity of dealing with a phenomenon in all its complexity. To make an analogy with the social sciences, it is as if a political scientist worked out a means of predicting the exact electoral vote of a presidential candidate without being able to tell which way any particular state would go.

The law that Descartes used to analyze the simple problem of the collision of two bodies was that of the *conservation of momentum*. Momentum, to Descartes, was the product of the weight of a moving body and its velocity. Newton later made the minor but significant substitution of *mass* for weight, a distinction which need not concern us at this point. The law asserts that when two bodies collide, the sum of their momenta will not change. It is best illustrated by means of an example. (See Fig. 2-3.)

Imagine two bodies free to move on one of the "frictionless" surfaces beloved to writers of physics texts. One is stationary and weighs 3 kilograms (kg), while the other is moving at 10 m/s and weighs 2 kg. (From this point on we shall in most instances employ the highly rational metric system of units, eschewing the English system, which is after all merely a modern patchwork codification of several sets of nearly unrelated medieval trade units, designed to minimize the social dislocation involved in adopting new units, a practice the French might call typically Anglo-Saxon.) Before collision, the total momentum (in kilogram-meters per second) is

$$p = (10 \times 2) + (0 \times 3) = 20 \text{ kg} \cdot \text{m/s}$$

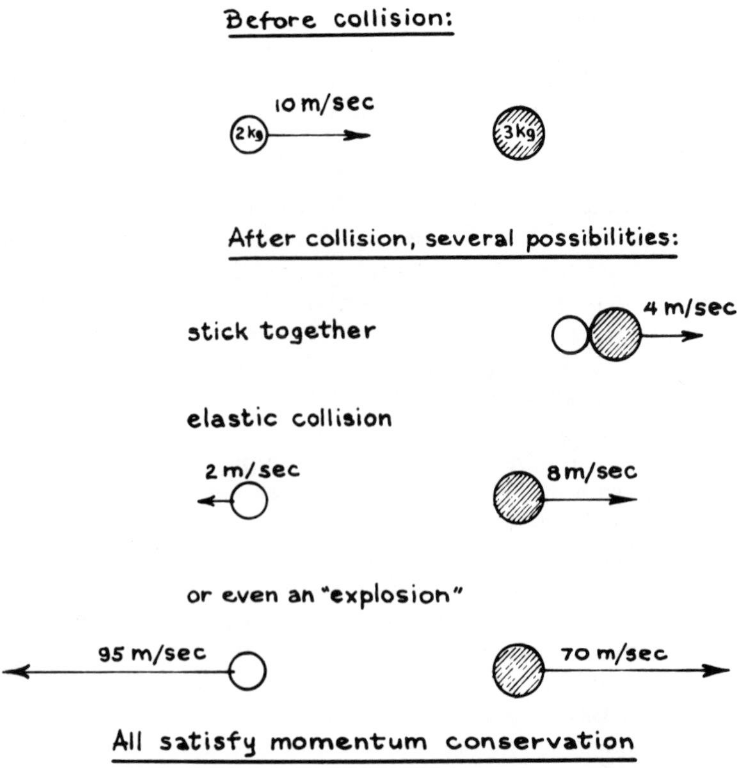

Before collision:

10 m/sec

(2 kg) ⟶ (3 kg)

After collision, several possibilities:

stick together

4 m/sec

elastic collision

2 m/sec   8 m/sec

or even an "explosion"

95 m/sec   70 m/sec

All satisfy momentum conservation

Figure 2-3

The use of p for momentum is another peculiar tradition, and the unit kilogram-meters per second unfortunately has no name of its own.

The law says that after the collision this sum will be the same. That is all it says. It does not pretend to assert what speed either body will have. Further information is necessary to settle that question.

The simplest case comes if the additional information is merely the qualitative assertion that the bodies get stuck to each other. Then we have a combined body of weight 5 kg. In order to have a momentum of 20 kg-m/s, the same as before the collision, this body must have a speed of

$$20 \div 5 = 4 \text{ m/s}$$

To give another example, the additional information could be a measurement of the speed of one body after the collision. For example, we might find that the struck body is moving at 8 m/s after the collision, in the original direction of motion of the moving body. Its momentum is then

$$3 \times 8 = 24 \text{ kg} \cdot \text{m/s}$$

What are we to make of this curious situation? We find we actually have more momentum than we started with! Descartes himself was baffled by this example and concluded that when a light object strikes a heavier one, it must recoil without budging it one iota, a conclusion that flies in the face of common sense. Here Huygens came to the rescue. He realized that momentum must not merely take into account the speed of motion but also its direction. Motions in opposite directions cancel each other. If we count a body moving to the right as having positive momentum, one moving to the left must have negative momentum.

Thus the 2-kg body must have momentum −4 kg-m/s. To find its velocity, we divide the momentum by the mass, getting −4/2 = −2 m/s.

This example was not chosen at random. Note that after the collision the larger ball is moving 8 m/s to the right and the lighter one is moving 2 m/s to the left. Thus they are moving apart with a relative speed of 10 m/s. This is the same as their speed of approach, before the collision. When this happens, the collision is referred to as *elastic*. Its significance will become apparent later when we introduce yet another conservation law, that of energy.

The law of momentum conservation also covers the situation in which the 2-kg body is moving 95 m/s backward while its partner moves 70 m/s forward, and so on through any combination that gives the right total momentum.

Indeed, we have already oversimplified the problem by assuming that the collision is head on and that the bodies move along the original line of motion after collision. If we consider motion in two dimensions, like the balls on a billiard table, a wide variety of possibilities is opened, and both the analysis and the amount of information needed to describe the process completely become more complex. And if we were to go into the question of what is going on during the instant the two bodies are actually in contact, we would have to deal with a highly complex motion. These are among the many details the law has chosen to sweep under the rug.

But in this very incompleteness lies the value of the law. A whole host of different processes is covered by the same quantitative statement. Though it describes no single process in detail, it permits us, for example, to tell exactly what happens to one body after the collision by means of measurements made solely on its partner, which is itself no mean accomplishment.

AN AMUSING RESULT: THE CENTER OF MASS DOESN'T MOVE

To close this chapter, we derive an amusing side result of the law of momentum conservation. It is of no great significance in itself, but it

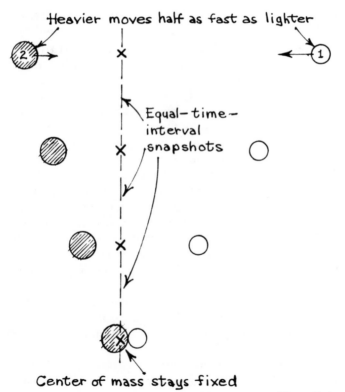

Heavier moves half as fast as lighter

Equal-time-interval snapshots

Center of mass stays fixed

Figure 2-4

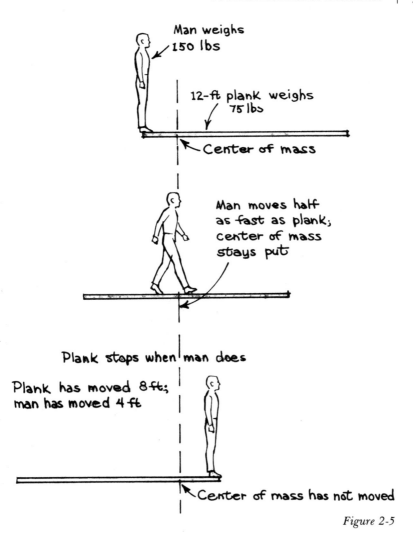

Man weighs
150 lbs

12-ft plank weighs
75 lbs

Center of mass

Man moves half
as fast as plank;
center of mass
stays put

Plank stops when man does

Plank has moved 8 ft;
man has moved 4 ft

Center of mass has not moved

Figure 2-5

provides a good example of the application of the law. Furthermore, it will prove useful later in the development of the theory of relativity. The theorem in question is the statement that if the center of mass of a group of objects is stationary, no interaction among the objects can cause it to move.

The concept of *center of mass* was familiar to the ancient Greeks and is equally familiar to any child who has played on a seesaw, for the center of mass is nothing but a modern term for the balance point. Few readers are unaware that if a 90-lb child wishes to balance his 45-lb younger sister on a seesaw, he must sit half as far from the center as she does. Stated somewhat more mathematically, a seesaw is in balance if the product of *weight* and *distance* is the same for both riders. A child weighing 10 percent more need be only 10 percent closer to the center than his partner; if he weighs twice as much, he must sit half as far out.

If we consider the somewhat more difficult case of two *moving* objects, it is clear that if the center of mass is to stay put, the heavier must move more slowly than the lighter. If a 90-lb object and a 45-lb one are approaching each other, as long as the heavier one moves half as fast, the center of mass, the balance point, stays put.

But from Huygens' point of view, this is merely the situation where the *total momentum is zero*. Two objects approach (or recede from) each other, the heavier at the proportionally slower speed. Their momenta remain equal and opposite. As long as no external influences come into play, the total momentum will remain zero.

Thus, if two cannonballs are fired at each other and one is twice as heavy but moving half as fast, if they stick together they will be at a dead halt; and at all times in their motion, the center of mass will be two-thirds of the way between, closer to the heavier object, as shown in Fig. 2-4.

As another example, imagine a tug-of-war between two people on roller skates. The heavier will move less, and they will ultimately collide at their center of mass, closer to the heavier one.

As a final example, consider Fig. 2-5, in which a 150-lb man is standing at one end of a 75-lb plank set on the ice so that it moves freely. As he starts to walk to the other end, he pushes the plank back. Since it weighs half as much, momentum conservation dictates it must move twice as fast. If the plank is 12 ft long, the man actually moves 4 ft while the plank slides back 8. When he gets to the other end and stops, the plank stops too. Since he has moved half as far as the plank, which weighs half as much, the center of mass has not moved.

CHAPTER THREE

The Denouement: Newton's Laws

---

*He has so clearly laid open and set before our eyes
the most beautiful frame of the System of the World,
that if King Alphonse were now alive he would not
complain for want of the graces of simplicity or of
harmony in it.*

—ROBERT COATES, PREFACE TO THE PRINCIPIA
second edition, 1713

*I*N 1665, A SCANT 33 years after the publication of
Two New Sciences, a young Fellow of Trinity College at Cambridge sat in a
farmhouse in the quiet Lincolnshire village of Woolsthorpe and put the
finishing touches on Galileo's "vast and most excellent science." Driven
from the crowded university town by the Great Plague, Isaac Newton
made remarkable use of his period of forced isolation from the intellectual environment he had adopted, returning to the small freehold on
which he had been born only 23 years before.

A comparison of Galileo and Newton provides an interesting study
in contrasts. Galileo's worldliness and arrogance could hardly be more
remote from Newton's polite and almost mystical reserve. Galileo
thought on his feet and was fond of public controversy; Newton was
moody and introspective and left his friends to fight most of his battles
for him. Where the former could hide his skepticism behind a formal
capitulation to the Inquisition without unduly burdening his conscience,
the latter remained throughout his life a convinced, if not fanatical,
Christian. Galileo's family hoped his studies would restore him to the
elevated rank they regarded as the proper family station in life, while it

seems likely that Newton's parents would have been content to see him succeed them as a simple dirt farmer.

Young Isaac, however, seemed to prefer reading to farm chores, so his family concluded that it might be wiser to send him to Cambridge University, where several of his relatives had been educated. He arrived there to find that years of civil war had left the university a shambles, both physically and intellectually. Few of the professors were capable of offering much in the way of intellectual stimulation. Furthermore, unlike the great Italian universities, Cambridge had little use for natural philosophy, Newton's primary interest. The chair in this subject was occupied by Isaac Barrow, an amateur mathematician whose first love was theology. Barrow had no illusions about his own qualifications. He tutored Newton as best as he could and willingly relinquished his position as soon as Newton was mature enough, at 26, to take his place.

Even for a natural recluse like Newton, the isolation of a scientist at Cambridge was almost unbearable. He might have been better off in London, among companions who were passionately dedicated to science, such as the astronomer Edmund Halley, the architect Christopher Wren, and the philosopher John Locke, as well as a rival physicist, Robert Hooke.

The members of this circle shared a world view in which there was no room for the supernatural or any limit to the power of reason to unravel the mystery of the cosmos. Their God might have laid down the laws of nature at the creation, but He did not intervene in the daily affairs of His world. They distrusted the philosophy of Descartes, which looked to them like an attempt to create a new scholasticism cloaked in the garb of science in order to reassert the intellectual authority of the Church of Rome. They saw in Newton a potential champion who could meet the cartesians on their own ground.

Newton had little heart for any such struggle. He was painfully sensitive to the slightest criticism of his work, even on purely technical grounds. It took all of Halley's persuasive powers to coax him into producing his greatest work, the *Philosophiae Naturalis Principia Mathematics*, (commonly known as the *Principia*), published in 1686.

NEWTON'S LAWS

Newton's complete theory of motion, one so perfect that it remains nearly intact to this day, was based in the main on two key insights:

1. The central problem in mechanics is that of change of state of motion, i.e., deviation from the behavior described in the principle of inertia.

2. Such a deviation can result only from the interaction of two objects, both of which have their motion altered in the process.

In order to treat the problem quantitatively, Newton needed three concepts—acceleration, force, and mass. Acceleration, as defined originally by Galileo, he took as the quantitative measure of the rate of change of state of motion. Force, as a measure of the strength of the agent causing the change, was a concept already prevalent in his time. Mass was in great part in his own invention, an index of the capacity of a body to resist a change in its state of motion. Newton regarded it, in fact, as the true measure of the "quantity of matter" in a body. As such, it is akin to the common meaning of the word *weight*. The distinction between weight and mass will be discussed later in this chapter.

Newton adopted momentum, in Huygens' directional version, as the true measure of motion, substituting mass for weight in its definition. What he added was a scheme in which the transfer of momentum from one body to another was a smooth continuous flow, rather than something which could only be analyzed in the before-and-after fashion of the preceding chapter. This scheme was expressed in three laws, which we present in Newton's own words:

LAW I: Every body continues in its state of rest, or of uniform motion in a right [straight] line, unless it is compelled to change that state by a force impressed on it.

LAW II: The change in motion [rate of change of momentum] is proportional to the motive force impressed; and is made in the direction of the right line in which that force is impressed.

LAW III: To every action there is always opposed an equal reaction; or, the mutual actions of two bodies are always equal, and directed to contrary parts.

The meaning is somewhat obscured by Newton's choice of language, which was designed to appeal to an age in which Euclid's geometry remained the finest intellectual monument. The first law we recognize as Galileo's principle of inertia. The second is usually restated as the equation

$$F = ma \qquad (3\text{-}1)$$

because in most situations the mass of an object does not change. Thus any change in momentum must involve a change in velocity only. The rate of this change is the acceleration. The rate of change in mass times velocity is then mass times acceleration.

The metric unit of force is defined by Eq. (3-1). It is the force required to give a mass of one kilogram an acceleration of one meter per second squared. It is called the *newton* (N), following the practice whereby physicists enshrine the memory of certain departed colleagues in the name of some appropriate unit.

The third law is easy to understand if we translate "action" as "change of momentum." Whatever one body gains, the other must lose. It is simply the law of momentum conservation.

THE CRUCIAL PEG

At first glance, Newton has added very little to the work of Huygens. But that little bit is crucial, for it is the peg on which he hung a complete theory of motion.

Newton treats the transfer of momentum from one body to another as a gradual process, and gives the name *force* to the rate at which it occurs. The power of his approach lies in the possibility of discovering laws of force which will enable us to predict in advance what forces will come into play when two bodies interact.

For example, the force of air resistance on a ball can be obtained from a formula that involves the size of the ball and how fast it is moving; the force exerted by a spring depends on how much it is compressed or stretched. From Newton's day on, the study of motion was reduced to a search for formulas of this sort. Once they are found, every detail of the motion can be predicted by use of the second law.

In the *Principia*, however, one force law was more important than all the others. This was Newton's *law of gravity*, which he used to explain the motions of the moon and the planets in terms of the same force that causes objects to fall on earth. This was his unanswerable challenge to the cartesians. Before we study it, however, it is necessary to take a look at how circular motion fits into the newtonian scheme.

GOING AROUND IN CIRCLES

The task of this section is to persuade you that motion in a circle at constant speed is in fact a form of accelerated motion, requiring a force that points toward the center of the circle. This result was yet another of Huygens' relatively unsung contributions, although Newton apparently got the same result independently, some time later.

The argument begins with an appeal to your intuition that it does take a force of some sort to change the direction in which an object is moving and that the force must be perpendicular to the motion. The next step is to show that exactly such a force is present when an object moves in a circle. We will then show a kinship between circular motion and more ordinary forms of acceleration. Finally, we will obtain a formula for this acceleration.

Imagine a large, heavy ball rolling slowly across a level floor. If we get behind the ball and push in the direction it is moving, it will speed up. Similarly, if we get in front of the ball and give a push opposite to its motion, it will slow down.

To change the direction in which the ball is moving, however, takes a push from the side. If the push is at all in the forward direction, the ball will speed up as well as change directions; if it is at all from the front, the ball will slow down. Only a push at right angles to the motion can change the direction without also changing the speed.

Now let the ball (still on the floor) be attached to a rope, the other end of which is tied to a post. The ball will move in a circle. There is certainly a force being exerted in this case; the rope is taut, and the post must be firmly anchored to the floor. The only direction in which the ball can move is perpendicular to the rope, maintaining a constant distance from the post. The rope can pull in one direction, toward the post. If we ignore friction, so that the ball does not slow down, the strain on the rope remains constant. Thus we have a force on the ball that is constant in strength but continually changes direction, always pointing to the center of the circle.

Suppose that we place a bright light on the floor, far from the circle, and observe the shadow of the ball on the opposite wall, as shown in Fig. 3-1. It moves back and forth in pendulum fashion, continually slowing down, reversing, and speeding up again. Here we have applied the principle of superposition to look at the motion of the ball in one dimension only, ignoring its motion toward or away from the light. We find that we have an obvious example of accelerated motion. We can imagine the motion of the ball to be compounded of two such motions at right angles to one another.

Each time the ball travels halfway around the circle, it has reversed direction, exactly as if it had been brought to a halt and then speeded up in the opposite direction. If we call its velocity then v, it must have been $-v$ when it started out on the opposite side of the circle. The change in velocity is $2v$, as in the case of a ball thrown vertically upward, which returns to the ground with its velocity reversed. The faster the ball is moving, the bigger the change in velocity and hence the bigger the acceleration. The faster the motion, the greater the tension in the rope.

Now that we know the change in velocity of the ball, it is possible to find the average acceleration for a trip halfway around the circle. To do so, we must also know how long the whole process takes. That depends on how far the ball went and how fast it was going. The circumference of

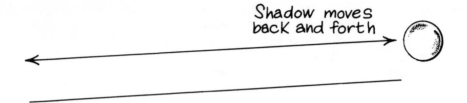

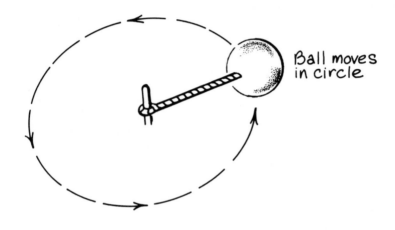

Figure 3-1

a circle of radius r is $2\pi r$, and the ball traveled half this distance. The time required is the distance divided by the velocity, $t = \pi r/v$. Using the definition of acceleration, we then have

$$a = \frac{\text{change in velocity}}{\text{time}} = \frac{2v}{\pi r/v} = \frac{0.64v^2}{r}$$

We must remember, however, that this is only the average acceleration, the end result of a force that has pulled in a variety of different directions as the ball went through its half turn. In fact at the end, the force is exactly opposite its direction at the beginning. The instantaneous acceleration therefore must surely be greater. Calculating it, however, requires some mathematical concepts that are not otherwise needed in this book, so we simply present the result:

$$a = \frac{v^2}{r} \tag{3-2}$$

The commonsense significance of this formula is not hard to see. The r in the denominator simply means that it takes more force to hold an object in a small circle than in a large one because it gets around faster. Thirty miles per hour may be a placid speed on a freeway interchange, but if you round a city street corner at this speed, you will hear a great squealing of tires. They are squealing because they do not like to handle a force perpendicular to the direction they are rolling in. The v in the numerator is squared because it makes the situation worse in two different ways: (1) there is more velocity to change; (2) the change takes place more rapidly. A curve that is safe at 60 mi/h is not twice but 4 times larger than one that is safe at 30 mi/h.

As a final example of the application of this formula, consider the path taken by a racing car rounding a curve, as shown in Fig. 3-2. A good racing driver enters the turn in the outside lane, moves to the inside

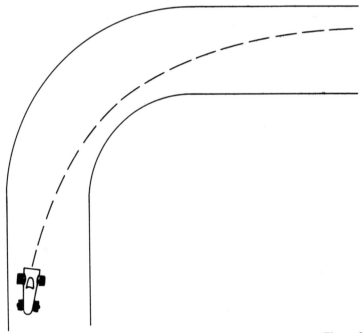

Figure 3-2

midway through the turn, and winds up on the outside. This path has the largest possible radius and thus permits the car to maintain the highest possible speed.

If it bothers you to think of circular motion at constant speed as accelerated, it is probably just another case of the confusion generated by using words one way in physics and another way in ordinary speech, where the word *acceleration* definitely implies that something is going faster. In physics, it means any deviation from inertial motion. We have already seen how a simple change of sign handles the problem of slowing down. What we have done in this section is merely to move one step farther from the common-language use of the word acceleration.

FALLING BODIES IN NEWTON'S SCHEME

Surprising as it may seem, Galileo's simple description of falling-body motion proved awkward to fit into Newton's grand scheme. But, making a virtue of necessity, Newton used the resolution of this problem as the starting point for his law of gravity.

To start with, since a falling body undergoes accelerated motion, Newton was forced to conclude that there must be a force acting on it—a very peculiar force, indeed, because since the acceleration is constant, the force is constant. Wherever the body is, whatever its speed, this force apparently does not change.

Furthermore, the acceleration is the same for all falling bodies. This is a most astonishing feature. There is only one way to account for it: the force must be proportional to the mass of the object. Then, in comparing two objects, if one is twice as massive as the other it will also have twice the force on it. The two effects compensate, and the acceleration remains the same.

This is the reason for the necessity of making a distinction between the *mass* of an object and its *weight*. The latter is our mysterious force that causes bodies to fall. By some fantastic coincidence, it seems to be proportional to the mass, but even in the absence of this force or when responding to other forces or moving horizontally, a body still displays mass. Indeed, we find, when sufficiently careful measurements are made, that the *weight* of a body is slightly different in different parts of the world but its *mass* never changes. Newton correctly emphasized the distinction between the intrinsic, universal property of *mass* and the peculiar phenomenon of a force related to mass, *weight*.

Numerically, the weight of an object is its mass times the acceleration due to gravity:

$$w = mg$$

Thus, an object that has a mass of 1 kg *weighs* about 9.8 N, according to a physicist. This causes some confusion between the language of physics and common usage, where mass units (gram, kilogram) are referred to as *weight*. Of course, what is being referred to in common parlance as weight is most nearly akin to the physicist's concept of mass. Just because 1 kg of gold weighs somewhat less at the equator than at the pole does not mean there is any less gold there.

The coincidence of a force exactly proportional to mass disturbed physicists for generations after Newton. In our own century, it disturbed Albert Einstein so much that it led him to abandon newtonian physics altogether; but that story must be left for Chap. 12.

THE NEWTON CULT

The publication of the *Principia* put an end to Newton's isolation, bringing him instant fame. More and more he found himself drawn to London, sometimes as an official spokesman for his university. He found the stimulating city life much to his liking and within a few years abandoned Cambridge for good. He was given the lucrative post of master of the Royal Mint. Though the job was intended as a sinecure, Newton took his work seriously. He began the practice of fluting the edges of coins to prevent the precious metals from being shaved off and personally attended the hangings of counterfeiters.

Most of all, however, Newton was on display, an intellectual adornment for a Britain reborn after decades of civil and religious strife. His London home became the center of social life for British and visiting foreign scientists. In this glittering public life, he grew fashionably corpulent and took a lively interest in genealogy, searching in vain for noble roots for the modest Newton family tree. Of Newton the Cambridge recluse but one vestige remained; he never married and, as far as generations of historians have been able to uncover, remained a lifelong celibate.

Once settled in London, Newton abandoned all serious scientific work. Privately, however, he indulged in theological and mystical speculations that would have outraged a younger generation that had anointed him the saint of reason.

After his death in 1727, the Newton legend continued to grow. The cult was nurtured by the otherwise skeptical thinkers of the Enlightenment. The coming of the industrial revolution brought new, more practical concepts ill suited to the newtonian framework. But by then his intellectual authority was so overwhelming that all innovations were quickly tamed by rephrasing them in ways that emphasize their connections to Newton's laws. This peculiar bias is retained in most introductory physics texts to this day.

The Moon is Falling

He lives below the senseless stars and writes his meanings in them.

—THOMAS WOLFE

HERE WERE GOOD reasons why the law of gravity created such a stir. First and foremost, it was an idea whose time had come. The free-thinking philosophers of Newton's era were eager for a simple, natural explanation of the motions of heavenly bodies. The "martyrdom" of Galileo had only served to strengthen their resolve to expunge the supernatural from the skies.

Another contributing factor was that, in Newton's time, astronomy was regarded as a high-priority, practical science. For a maritime nation like Britain, whose merchantmen ranged the seas of the globe, improved safety of navigation was regarded as a very urgent matter. The Royal Greenwich Observatory was built specifically to improve celestial navigation. For the first time in history, astronomers had been liberated from casting horoscopes to make a living, and they were pleased to have a meal ticket more in keeping with the spirit of their time. To have a rational physical theory of the motions of the moon and planets helped elevate their science to its new status.

THE LAW OF UNIVERSAL GRAVITATION

The best way to grasp the significance of the law of universal gravitation is first to state it and then analyze it in detail. In Newton's words, all bodies are attracted toward one another by a force proportional to the product of their masses and also inversely proportional to the square of the distance separating them. In mathematical language

$$F = \mathfrak{g}\,\frac{Mm}{R^2} \qquad (4\text{-}1)$$

where M and m are the masses of the objects concerned, R is the distance separating them,* and $\mathfrak{g}$ is a universal constant that determines the strength of the force.

The beginning of the chain of reasoning leading to this law has already been outlined at the end of Chap. 3. Newton was forced from the outset, in order to explain the motion of falling bodies, to postulate a force proportional to the mass of the falling body. By what means could such a force act? After Galileo had argued convincingly that falling bodies moved as well, indeed, even more perfectly, in a vacuum as in the air, Newton could hardly revert to earlier ideas and ascribe this action to the medium. Thus Newton's thought was directed inevitably to the concept of an *action at a distance,* an action that takes place between two objects not in contact. To most of his scientific peers, and indeed to Newton himself, this was the most disturbing feature of the law of gravity. But, building on the achievements of Galileo, he hardly had a choice.

The next problem is uncovering the identity of the "other body" demanded by the third law. Again, there appears one most logical candidate. Once we visualize our spinning ball of an earth, with "down" miraculously always toward its center, it becomes the obvious choice.

At this point, the third law intervenes decisively. It makes no distinction between the earth and a falling apple. The force on each is the same. The immense disparity of their masses is what is responsible for the fact that the apple falls, while the earth moves only imperceptibly. Then why make the distinction between the two bodies in the law of force itself? If the force must of necessity be proportional to the mass of the apple, why is it not also proportional to the mass of the earth?

This is not a question of logical necessity; it is one of style. The universal laws of motion, which apply to all objects, all forces, make no distinction between the two partners in the interaction given the name *force.* But there is no logical reason why some *particular* force cannot make this distinction. For example, when a spring pushes on a ball, it is the characteristics of the *spring,* not the *ball,* that determine the force between them, though the force is the same on each and their individual reactions to it are determined solely by their masses. Why not a similar

---

* One might legitimately ask what is meant by distance of separation for objects of extensive size. Newton was able to show by difficult geometric arguments that for spherical objects, at least, a mass acts as if concentrated at its own center.

distinction in the case of gravity? Newton chose another route—to preserve the uniqueness of his system, the beauty of a perfectly symmetric theory—by introducing the mass of the earth, at the time unmeasurable, as a variable in his theory.

Besides, a theory ignoring the mass of the earth would be unaesthetic; if the force of gravity depends only on the arbitrary mass of the particular falling body, it is the *earth itself* that has a passive role, hardly the appropriate choice if one must choose active and passive partners. Once again, as so many times before and since, the search for harmony, for beauty, was a guiding force in the choice of a hypothesis. But aesthetic considerations are not enough; the theory must also explain the facts.

The numerator of Eq. (4-1), the term Mm, is at this stage of the argument completely determined. For the only way a force can at the same time be proportional to the mass of the earth and to that of a falling body is to be proportional to the product of the two.

None of these arguments, of course, lent any support to Newton's claim that this force was not confined to the earth but was shared by all bodies, including the sun and planets, or that it diminished as the square of the distance between objects. To prove this, Newton made use of the detailed laws of planetary motion, discovered a generation earlier by Johannes Kepler. The story behind these laws is one of the most unusual in the history of science.

THE MANGY DOG AND THE MAN WITH THE GOLDEN NOSE

Johannes Kepler was born 8 years after Galileo, in the obscure south German town of Weil der Stadt, the son of a soldier of fortune and an innkeeper's daughter. That one so born could obtain an education was a tribute to the good sense of the Dukes of Württemberg, who provided scholarships for students of slender means, a practice by no means common at the time.

His professors at the University of Tübingen saw in Kepler a personality too passionate and unsubtle to survive as a clergyman in an era of religious strife and encouraged his interest in astronomy and mathematics. Just before he completed his studies for the ministry, a post as mathematics teacher in the Lutheran school at Graz in Catholic Austria came vacant, and Kepler was persuaded to accept it. His pay was meager, but his duties were undemanding, since the number of students with any interest in the subject was negligible. He was thus free to pursue his astronomical studies, and his published writings on the subject soon earned him a modest reputation throughout central Europe.

Though rewarded by fame, Kepler was never to know either prosperity or peace of mind. Tortured by real and imagined illnesses, poverty, and religious persecution, his life was a frantic juggling act to keep the wolf from his door and the demons in his mind at bay. He chose for himself the metaphor of a mangy dog.

The crucial event in Kepler's life came in 1600, when a ban on Protestants forced him to flee Graz for Prague, where he was offered refuge as an assistant to Tycho Brahe, the foremost astronomer of his time. Tycho had deserted his native Denmark for rather different reasons.

Tycho's origins were as splendid as Kepler's were mean. Born to one of the leading families of Danish nobility, he was pointed by family tradition toward a career in the service of the crown, but when he was a 14-year-old student at the University of Copenhagen, an incident seduced him from this path. The incident was a total eclipse of the sun, always an awesome and terrifying sight. But what most impressed Tycho was the fact that it had been predicted, with seemingly uncanny precision, by astronomers. Tycho decided that any calling capable of such a feat was well worth devoting one's life to.

Studying astronomy on the sly while supposedly preparing for a legal career, Tycho soon discovered that the precision that had attracted him to the subject still left a great deal to be desired; in particular, tables of planetary motion could be trusted only for a few decades or so. After this, they were likely to be off by days or even weeks. He correctly perceived that what was needed were better instruments, and he combed northern Europe in search of artisans who could build them.

His big break came in 1572, in his twenty-fifth year, when what is now known as a *supernova* erupted in the northern sky. With his superior instruments, he was the only astronomer to show that this bright object was far beyond the atmosphere or even the planets, in the domain of the supposedly unchanging stars, a blow to the scholastic cosmology. This discovery at once made him one of the most celebrated astronomers in Europe.

Never before had Denmark produced a philosopher of such reknown, and King Frederick II was determined not to lose him to more cultured lands to the south. He made Tycho an offer unprecedented in the history of science: the island of Hveen, as a site for his observatory, and generous grants from the state treasury, to maintain it and fill it with the finest instruments skilled hands could fashion.

This was "big science," even by present-day standards. Tycho presided over a large staff of artisans and students, with duplicate equipment that permitted four simultaneous independent observations, all but eliminating human error. Tycho and his students improved the precision of astronomy, frozen at 10 minutes of arc for 15 centuries, by a factor 10. All this was done with the naked eye, for the astronomical telescope was still two generations in the future.

But Tycho was far from a hero to some of his countrymen. A great bear of a man, fully endowed with the arrogance of his class, he upbraided his peers for their preoccupation with hunting, gluttony, lechery, and dueling (Tycho himself, in his youth, lost most of his nose in a duel, replacing it with one fashioned from an alloy of silver and gold). He further scandalized them by marrying the daughter of a peasant on one of the Brahe estates, which by Danish custom made his children illegitimate.

Frederick's successor, Christian IV, had little patience with the impetuous astronomer. Besides, he was strapped for funds due to military misadventures. Citing as excuse Tycho's exactions of labor from the peasants of Hveen, which were undeniably excessive, Christian deprived his observatory of much of its income. Incensed, Tycho left Denmark to enter the service of Rudolf II, the Holy Roman Emperor, who was himself an amateur astronomer.

Tycho carried to Prague his instruments and his precious tables of observations. He hoped to crown his fame with a new, more precise version of the earth-centered cosmology of the Greek astronomer Ptolemy of Alexandria, which had endured for 14 centuries. Kepler was hired to carry out the arduous calculations required to complete this task. Confident of his own powers of persuasion, Tycho cared little that his new assistant was a convinced copernican.

THE TWO GREAT RIVAL COSMOLOGIES

The motions of the planets have been a puzzle since the very dawn of astronomy, which in many civilizations predated the invention of writing. The stars appear to wheel around the earth in unison, as if attached to an immense, rotating sphere. The sun and moon follow a circular path through the starry sphere, completing their journeys in a year and a month, respectively. The planets, however, are mavericks.

Mercury and Venus swing back and forth across the sun's disk, following it in its journey through the skies. Mars, Jupiter, and Saturn trace predictable paths, but their progress is fitful. From time to time, they seem to reverse direction for a few months. The three remaining planets were not discovered until the era of the telescope.

The idea that the earth is itself a planet and like the others circles the sun is nearly as old as astronomy itself, for it offers a ready explanation of these observations. The orbits of Mercury and Venus lie inside ours, so we never see them far from the sun. Mars, Jupiter, and Saturn lie outside and move slower than the earth. When the earth overtakes one of these slow voyagers, it will seem to be moving backward. This theory was first set in writing by Aristarchus of Samos in the fourth century B.C.

Ptolemy recognized the virtues of Aristarchus' system, but the notion that the earth could move without our being aware of it seemed to fly in the face of both common sense and the physics of his time. He developed a quite accurate scheme in which the planets moved on *epicycles,* circles whose centers in turn moved on other circles. This system was refined by the great Arab astronomers. The final improvement was due to Tycho, who incorporated some of the advantages of the sun-centered scheme by centering all the planetary epicycles on the sun, as shown in Fig. 4-1.

It is important to realize that the issue between the sun-centered and earth-centered systems was never one that could be settled by

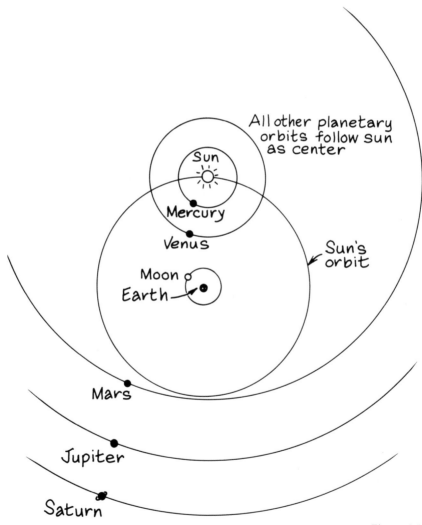

Figure 4-1

astronomical observations, which only show where the planets *appear* to be as seen from earth. Ptolemy's scheme actually gave as accurate a picture as that of Copernicus and was only slightly more complicated. Copernicus was forced to resort to small epicycles, for the orbits of the planets around the sun are not perfect circles.

If we were to use Fig. 4-1 as the blueprint for a mechanical model of the solar system, it would serve equally well for the copernican system. The only difference would be the choice of which body to attach to the stand, the sun or the earth. The choice comes down to the question: Is it reasonable to regard the earth as moving? This is a question in physics, not astronomy. Ptolemy, Tycho, and Kepler all realized this full well, and each based his position on his own physical intuition.

KEPLER CHARTS THE MOTIONS OF THE PLANETS

Tycho was to live only one short year after Kepler's arrival in Prague. Kepler inherited both Tycho's job and his notebooks, though his rights to the latter were challenged by Tycho's heirs. What he did with them was a tour de force of data analysis, one that looks impressive even from the vantage point of the computer age. It reveals a sensitivity to both the value and the limitations of precision measurement that was centuries ahead of its time.

The record of Kepler's labors survives in his two great works, *Astronomia Nova* and *Harmonice Mundi*. They are unique in the annals of science in presenting not only his conclusions but a complete account of the tortuous path he took to them, filled with false starts, blind alleys, and wrong hypotheses discarded only after months of laborious hand calculations. Interspersed are poems and fragments of verse in which he castigates himself mercilessly for his temporary failures and exults wildly in his final triumphs.

Most of all, however, Kepler reveals the mystical vision that drove him to this great rationalistic triumph. He put the sun at the center of his universe because, as the giver of light and life, it was closer to God than the base earth and thus more worthy of the honor. He pursued the planetary orbits relentlessly, certain that, once known, they would provide a divine lesson in solid geometry and the laws of musical harmony. In this, the dominant passion of his life, Kepler was to fail. But on the way to this personal tragedy, he left behind three laws that endure to this day:

1. *The planets travel in ellipses with the sun at one focus.* Figure 4-2 explains the terms used to describe an ellipse, indicating the *focus*.

2. *The area swept out by a line drawn from the sun to a planet is the same in equal time intervals.* Figure 4-3 illustrates this. Each planet moves fastest

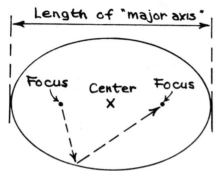

Figure 4-2 Structure of an ellipse. If it were lined with a mirror any light ray from one focus would reach the other, hence the term focus.

when it is nearest the sun, slowest when it is farthest away. If its farthest point is twice as far as its nearest point, for example, it will go half as fast at the one as at the other.

3. *The square of the length of each planet's year is proportional to the cube of the major axis of its orbit.* Figure 4-2 shows the *major axis,* the largest dimension of the ellipse. This law implies that the outer planets move more slowly in their orbits than the inner ones, and thus the length of the year increases more rapidly than the size of the orbit.

Though Kepler describes the orbits as ellipses, the orbits of the planets are very near to being circles. The laws are based on observation of almost imperceptible deviations from the simple behavior they would exhibit if the orbits were perfect circles. But these deviations are important, because they enabled Newton to demonstrate the inverse-square character of the force.

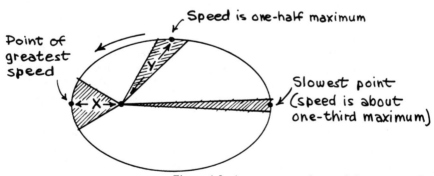

Figure 4-3 Areas swept out in equal times are equal.

NEWTON USES THE LAWS

A major portion of Newton's *Principia* is devoted to detailed, complex geometric arguments designed to explain the significance of these three laws. First, he showed that Kepler's second law is evidence that the planets move subject to a force directed toward the sun—*any* force of this type would produce the observed variation of velocity. For Newton's purposes, this was quite important, because it demonstrated that if the behavior of the planets can be accounted for in mechanical terms, i.e., by a force, it must be a force directed toward the sun. Newton then demonstrated that Kepler's first and third laws were possible *only* for a force that varied inversely as the square of the distance. To emphasize this point, he found the comparable laws for a number of other types of forces. For example, a force proportional to the distance would still produce elliptical orbits, but the sun would be at their center, rather than at a focus. And since such a force increases with distance, the outer planets would have to move faster than the inner ones to have greater acceleration, and all planets would turn out to have the same length of year.

It is interesting that Newton, after inventing the calculus, which enormously simplifies computations of the type required to interpret Kepler's laws, made no use of it in this connection. He fell back on intricate geometric arguments, and much of the *Principia* reads like a somewhat old-fashioned, very advanced high school geometry text. This was in response to the fashions of a time in which Euclid's geometry remained the highest achievement in mathematics. The use of his new methods might confuse his readers, and in any event they would have found arguments based on the calculus less convincing.

THE MOON IS THE LINK

But Kepler's laws spoke only of *relative* speeds of motion of different planets or of the same planet at different points in its orbit. Not knowing the magnitude of the force on the planets, knowing neither the mass of the sun nor the g in his formula, there was no way to complete the analysis; Newton had demonstrated that his law could account for the fact that Mars's year is nearly twice as long as the earth's, but the length of either year in itself could not be accounted for. Merely accounting for Kepler's laws could not destroy the dichotomy between terrestrial and celestial phenomena.

Fortunately for Newton, our earth has a moon. Since it is the same earth that is responsible for the acceleration of a falling stone and of the moon in its orbit, the only difference between the two phenomena is that the moon is farther from the center of the earth, and the resulting reduction in the force of gravity can easily be calculated. From the measurement of the acceleration of a stone on earth, he could calculate the acceleration the moon must have if subject to no force other than the earth's gravity. Would the calculation agree with the observed motion of the moon? Here was Newton's opportunity for a direct *quantitative* confirmation that the same force responsible for the stone's motion could account for the moon's.

Already in Newton's time the distance to the moon was known, for it is quite easy to measure. It is 380,000 km, which is 60 times as great as the earth's radius. Thus, the acceleration resulting from the earth's gravity, at the moon's orbit, must be less than that at the earth's surface by a factor of $(60)^2$, that is,

$$a = \frac{9.78 \text{ m/s}^2}{(60)^2} = \frac{9.78}{3600} = 0.00272 \text{ m/s}^2$$

If we assume the moon's motion to be simply that of a material object, with no force acting upon it other than the earth's gravity, this must be its acceleration as it travels in its orbit. This is Newton's "leap to the stars"—a prediction about the moon based on the behavior of falling bodies on the earth.

There now remains only to calculate the observed acceleration of the moon, using the formula obtained in the preceding chapter. The speed of the moon in its orbit is 1016 m/s.* Thus, its acceleration is

$$a = \frac{V^2}{r} = \frac{(1016)^2}{380,000,000} = 0.00271 \text{ m/s}$$

Considering the fact that the moon's orbit is not quite circular, but very slightly elliptical, the agreement is remarkably good.

Just to be sure we have not lost the train of the argument while worrying about the arithmetic, let us recap. We started by assuming that the moon is a material object subject to the same laws that apply on earth. We find that the acceleration of the moon is smaller than that of a falling body on the earth by exactly the square of the ratio of their distances from the center of the earth.

The result was electrifying. Of course, a single such success can turn out to be a coincidence, and final confirmation waited upon more detailed calculations of other phenomena. Yet this calculation was the clincher, a quantitative correlation between an earthbound and a

---

* This figure is obtained by the circumference of the moon's orbit, $2\pi \times 380,000$ km, converting it to meters, and dividing it by the length of the lunar month, 27.32 days, converted into seconds (86,400 s/day).

heavenly phenomenon. It provided the first link between a phenomenon on earth and one in the skies.

But the boldest aspect in the theory remained untested: its claim to represent a universal force that acted between all bodies proportional to the product of their masses. An inverse-square force exerted by the sun could be seen to account for the motions of the planets (including the earth itself). A similar force exerted by the earth could account for the motions of both falling bodies and the moon. Not knowing the masses of either the sun or the earth, Newton had feeble grounds for claiming the force was proportional to the mass of the large "central" object in each case. And he had no grounds whatsoever for assuming it would act between, for example, two rocks except for the negative evidence that since they are far smaller than the earth, the rocks would not exert on one another a force sufficient to measure. Nonetheless, so convincing were the theory's successes, and so well did it fit into the general spirit of newtonian physics, that few physicists seriously doubted the claim to universality from the date of its publication. It was not directly confirmed until a century and a half later, when Henry Cavendish developed an instrument to measure the feeble force between two objects in the laboratory. By then the faith in Newton's law of gravitation was so great that Cavendish did not describe his experiment as a "confirmation of the law of universal gravitation," but instead called it "weighing the earth," a title that rested on the also unproved claim that the force of the earth's gravity on a stone was proportional to the mass of the earth, as well as to that of the stone. Since he had measured the force between two objects both of known mass, Cavendish was the first to have all the information necessary to measure the constant g in Newton's formula. Once this constant is known, the weight of a known mass on earth can be used to calculate the mass of the earth, since all the other factors in the formula for the force on it are known.

For at least a century after the publication of the *Principia,* calculations of the motions of the planets remained the most important application and confirmation of both Newton's laws of mechanics and his law of gravity. It was not until well into the nineteenth century that the design of machinery, for example, ceased to be a trial-and-error process, and Newton's laws of motion received their first practical application.

HYPOTHESES NON FINGO

If the theory of gravity was Newton's greatest accomplishment, one feature of it was to prove his most severe embarrassment. That was the notion of action at a distance, that two bodies, though far apart, could exert a force on each other with nothing intervening but empty space. Privately, Newton dismissed this idea as absurd and was sure that there must be some material agent to transmit the force, though he had not the slightest clue to what it might be.

But his public stance was far different. With his almost paranoid distaste for the give and take of scientific debate, he opted out of the dispute with a haughty *"hypotheses non fingo,"* which translates roughly as "I do not engage in idle speculation."

Newton's posture on this issue was unfortunate for the subsequent history of physics. In his own time, it left the theory vulnerable to attack from the cartesians, who scented in it a "whiff of the brimstone," the very kind of occult explanation of nature that science was supposed to have banished forever.

The backlash, unfortunately, was worse than the attack. When later generations had resolved the struggle in Newton's favor, the belief in pure action at a distance became the supreme test of rationalism. When a better idea, the *field concept,* came along in the nineteenth century, it faced an uphill struggle against this prejudice. Fields will be introduced in Chap. 6 and will be a major theme of this book, for today the "agent" that Newton so cavalierly dismissed is considered to be fully as real as matter itself.

Indeed, the very mathematical form of the law of gravity fairly begs for such an explanation. The inverse-square law holds for any "influence" that radiates out equally in all directions from a central source, such as the light from a naked bulb. This is because the farther one gets from the source, the larger the total area over which it must spread. The total area increases as the square of the distance, so the fraction of the light that falls on any given area, such as a piece of paper, must diminish by the same power. If gravity is also something that spreads out in this fashion, it could scarcely escape obeying an inverse-square law.

Just such a line of reasoning led Hooke and Halley to guess that gravity would obey an inverse-square law, even before Newton settled the matter by proving that it led to Kepler's laws. Had Newton had more stomach for public disputes, or had he become less of a cult hero, this good idea might not have been set aside for a century and a half.

CHAPTER FIVE

The Romance of Energy

. . . may God us keep
From single vision, and Newton's sleep.

—William Blake

T HE CENTURY FOLLOWING the publication of the *Principia* was one of unbroken triumphs for newtonian physics and the astronomy it served so well. The calculus was honed into a fine analytic tool and the motions of the moon and planets charted to astonishing precision. Yet, toward the end of this period, quite a few scientists began to realize that from a practical standpoint, Newton's laws and the concept of momentum were only a beginning toward a science of motion, rather than the end they had seemed at first.

The problem was that newtonian mechanics took force as its starting point and went on from there. It offered no ready answer to the question that the pioneers of the industrial revolution confronted daily: What does it take to generate a force? Engineers and inventors wanted to create motion where none had existed before, and all Newton had to offer them was the assurance that however they managed to do it, an equal and opposite motion would inevitably arise in the process. This was some help, but not much.

At the same time poets and philosophers, who just a generation or so earlier had applauded the liberating influence of the newtonian spirit, began to speak of its darker side. The cool analytic method, seeking precision through a process of dissection, often lost sight of the beauty and unity of nature. While they still extolled the power of human reason, they feared a sterile rationalism that seemed to leave no room for the emotional wellsprings of creative thought.

The response of physics to these two criticisms, which arose from seemingly opposite poles, was a happy fusion in the concept of *energy*. Today energy ranks as the central unifying core of physics, one which links it to other sciences and to the practical world.

HOW DO WE "PAY" FOR A FORCE?

A few examples will serve to reveal the practical shortcomings of a strictly newtonian physics and point the way to new concepts that will make it more useful.

Consider first a bullet fired from a gun. Momentum conservation requires the gun to recoil with momentum equal and opposite to that of the bullet. Their combined momentum was zero before and remains zero afterwards. Yet simple common sense tells us that something significant has changed. Something has been taken out of a bit of gunpowder and transformed into motion of the bullet and gun. In the process, the powder has been transformed and lost the capacity to do this again. Somehow, our physics should be able to make a distinction between the situations before and after.

Next let us examine one of the most familiar practical uses of a force, to propel a vehicle such as a car. Much of the time, the car moves at a fairly constant speed, yet some force is still needed to overcome friction and air resistance. From the point of view of Newton's laws, this is a thoroughly uninteresting case. The motive force exactly balances the resistance, so the net force and acceleration are zero. The only recourse for the designer is to measure the force required and see to it that the motor is up to the task.

Newton's laws are not entirely useless in this situation. They do tell the designer how much additional force is needed to accelerate the car and how much force the brakes must exert to stop it. They also give a reminder that these forces produce equal and opposite reactions on the road.

But we know that the motor burns fuel, while the brakes do not. The brakes, however, do heat up, and the designer must find some way to get rid of this heat by maintaining a flow of air past the brakes. The question of how much fuel the motor must burn and how much heat the brakes must get rid of seemingly lies outside the whole science of motion.

When the car is rounding a curve at constant speed, the situation is entirely different. The driver need hardly exert any effort at all. But if the curve is taken at close to the maximum safe speed, the acceleration is nearly as great as in a fast start with the engine "flat out." Yet there is no significant increase in fuel consumption while rounding a curve. It is even possible to design the steering gear so that it takes zero driver effort to hold the car on the curve, but it would be unsafe; the car would have no natural tendency to straighten out when the steering wheel was released.

To speed a car up, we must take something from the fuel; to stop it, we must discard heat; to simply change direction, nothing need be taken from or lost to the outside world. Yet in each case, the force has about the same *strength*. It is the *direction* of the force that matters. A forward force must be paid for; one to the rear discards something; while a force perpendicular to the motion costs nothing.

Using these examples as a guide, we can write down a three-step program to develop a practical science of motion:

1. We need a new conservation law based on a nondirectional measure of motion, which unlike momentum does not cancel for motions in opposite directions.

2. The connection of this measure to newtonian physics must take into account the direction of the force relative to the motion, so that a forward force has a positive effect, a backward force a negative one, and a perpendicular force no effect at all.

3. Finally, we must find a connection with things that seem at first glance to have nothing to do with motion, such as heat or the power that resides in fuels and explosives, since motion does appear and disappear.

It is time to give a name to the mysterious something we are looking for, which manifests itself sometimes as motion and sometimes in other forms. It is called *energy*, a name borrowed by physics from the same poets and philosophers who deplored its sterility. We will begin our quest for energy by establishing its links to newtonian physics.

WORK AND KINETIC ENERGY

The first two steps are taken care of by expanding the newtonian language of motion. We introduce a new measure of motion, called *kinetic energy* (KE), which is defined by the formula

$$KE = \tfrac{1}{2}mv^2 \tag{5-1}$$

The $\tfrac{1}{2}$ in the formula comes from the same place as that in Galileo's formula for accelerated motion, as we shall shortly see.

The definition looks deceptively similar to that of momentum, but the fact that in this case the velocity is squared is a crucial difference. That is what makes kinetic energy a nondirectional measure of motion. Whether the velocity is positive or negative, its square is always positive. Motions in opposite directions do not cancel out.

The second step is to define a quantity called *work*, a measure of the change in energy due to the action of a force. Work is calculated by multiplying the force by the distance an object moves in the direction the force acts. For example, if the force is gravity, we count only the vertical distance moved. This definition is most easily put in mathematical form by using the cosine function from trigonometry:

$$W = Fs \cos \theta \qquad (5\text{-}2)$$

where θ is the angle between the force and the direction of motion. Even if you are unfamiliar with trigonometry, the meaning of the formula is easy to explain. If the force is in the forward direction, θ is less than 90°. For such angles, the cosine is positive. From 90 to 180° it is negative. At exactly 90° the cosine is zero, so no work is done. This is exactly what was demanded in step 2 above.

To illustrate the use of these concepts, consider an object that starts from rest, pushed by a constant force. It can only move in whatever direction the force pushes it. The angle between the force and the motion is then 0°, so the cosine is 1 and the work is simply Fs.

But since this is uniform acceleration, exactly the situation described in Chap. 1, we can use the formula $\frac{1}{2}at^2$ for the distance moved. We can also use Newton's second law and substitute ma for the force. Then we have

$$Fs = (ma)(\tfrac{1}{2}at^2) = \tfrac{1}{2}ma^2t^2 = \tfrac{1}{2}mv^2$$

This is offered simply as an example, for those unafraid of the mathematics, of how the work done equals the kinetic energy produced.

Do not be deceived by the preceding example: we are not simply looking at a new way to apply Newton's laws. To drive this point home, let us return to the example of a car moving at constant speed. In this case, the force exerted by the motor is still doing work, but this work does not go into increased kinetic energy because the force is opposed by friction and air resistance. The work, however, is not simply lost; it is producing other forms of energy, as indicated in step 3 of our program for a practical science of motion.

WHAT'S A WATT?

Because energy is a commodity in our civilization, and a terribly important one at that, it is worthwhile to pause briefly to take up the practical question of energy units.

The metric unit of energy is the *joule* (J). It is defined as the work done by a force of one newton operating over a distance of one meter. The joule, however, is a unit used mainly by scientists, because it is far too small for commercial use. The most common energy unit in commerce is based on the metric unit for *power,* the *watt.* (W).

Power is defined as the rate at which energy is converted from one form into another. If one joule of energy is used per second, the power is one watt. If you multiply the power by the time it is in use, you get back to energy. To measure electrical energy, it is common to use the hour as the unit of time. But even a watt-hour (Wh) is a terribly small amount of energy, so the unit in commercial use is the kilowatt-hour (kWh), or 1000 Wh. Since there are 3600 s in 1 h, 1 kWh is 36 million joules! Since this much energy costs only a few cents, delivered to your home, you can see what a small unit the joule is.

The *horsepower* (hp) is another familiar unit of power. In the metric system it is defined as 745 W. The horsepower rating of a car's engine represents the maximum it can put out. Unless you have a stick shift and are an unusually aggressive driver, you have never run the engine of a car at anywhere near its rating.

From time to time, unfortunately, you are likely to encounter something called the Btu, for British thermal unit. It is the energy required to heat up one pound of water by one degree Fahrenheit, and is used primarily to rate heating or cooling equipment. A Btu is roughly 1054 J. As the United States converts to the metric system, the Btu will hopefully become extinct.

THE MANY FACES OF ENERGY

The third step in the development of the broader energy concept begins with the recognition that when kinetic energy disappears, through the action of friction, heat is produced. Physicists began to realize this late in the eighteenth century. Who it was that first attempted to quantitatively relate the energy lost to the heat produced is subject to some debate, but certainly one of the more dramatic early efforts was that of a picaresque eighteenth-century American, Benjamin Thompson, better known under his acquired title of Count Rumford. Rumford was an energetic and ambitious Tory who fled America (deserting a wife and family) during the War of Independence. Entering a career in the British civil service, he pursued some rather practically oriented chemical studies that earned him appointment as a Fellow of the Royal Society only 3 years after his arrival in Britain. When a political turnover threatened his career, he decided to seek his fortune abroad, traveling extensively on the Continent. He settled for some years in Bavaria, where he played various political and military roles as an advisor in the court of the Elector Maximillian. It was from this sovereign that he received his title.

In the late 1790s, using the facilities of Maximillian's royal arsenal, Rumford employed a cannon-boring machine to obtain a crude quantitative estimate of how much heat is produced by expending a known amount of work.

This was one of the first steps toward a new science called *thermodynamics,* the study of the transformation of mechanical energy into heat and, of more practical interest, the reverse process. The industrial revolution was in full swing, and its symbol was the steam engine, which reverses Rumford's cannon-borer; heat is converted into mechanical energy. The practical demands of technology provided a strong motive for theoretical studies of this process. Much of the time, however, technology led science, rather than vice versa; practical men made improvements in steam engines, and the scientists came along later to explain why they worked. The relation of science to technology is a complex one, and the view that science is the starting point, with technology the follower, is a naïve one. The reverse is often the case. Even today, in an era when basic scientific research is fairly generously supported in nearly all the developed countries, it is still common for a new gadget to come on the market long before scientists fully understand the principles by which it operates.

Energy exists in, and can be transferred to, many other forms, for example, electrical and chemical. Thus, a complete law of energy conservation would read

$$\begin{bmatrix} \text{kinetic} \\ \text{energy} \end{bmatrix} + \begin{bmatrix} \text{chemical} \\ \text{energy} \end{bmatrix} + [\text{heat}] + \cdots + [\quad] + [\quad] = \text{const} \qquad (5\text{-}3)$$

In each of the brackets of this equation, the formula for energy is different. For heat, the formula involves the temperature, amount of material, and a constant called the *heat capacity,* which is related to the properties of the material containing the heat. Another example of one of the formulas that might be found in the brackets would be one relating the loudness of a sound, as recorded by a microphone, to the energy carried by the sound. For chemical energy, we simply measure the heat given off in a chemical reaction, such as the burning of gasoline.

This view of energy as some mysterious, chameleonlike entity that appears in a variety of guises but is never created or destroyed was very much in the spirit of a cultural movement of the late eighteenth and early nineteenth centuries. The musical and literary branches of the movement went by the name *romanticism,* which we will use for the movement as a whole.

The romantic movement was a reaction to the extreme rationalism of the enlightenment, to the sterile formalism of the music, art, and architecture of that period, and to the horrors of the new industrial society, with its "dark satanic mills." While it did not reject reason per se, the movement extolled the creative force of emotion and intuition. Romanticism had a scientific offshoot called *Naturphilosophie.* The exponents of this school came from all branches of science and put heavy stress on the basic unity of nature. Energy was their word for the vital principle behind all change, motion, growth, creativity, and passion. This idea was particularly popular in German-speaking countries; elsewhere, most scientists dismissed its supporters as fuzzy-minded dilettantes.

Without this sort of passionate conviction, the energy concept might never have come to be, for in its early stages of development the law of energy conservation had to be taken pretty much on blind faith. Most forms of energy were little understood, and quantitative tests were few and far between.

In the 1840s, however, most "respectable" scientists did an about-face and accepted energy wholeheartedly, though many boggled at the romantic connotations of the word and used *force* instead, which created some confusion. The turning point was a precision measurement of the conversion of mechanical energy into heat by the Scottish physicist James Prescott Joule, who had previously been one of the leading opponents of the theory. It is after him that the metric unit of energy was named.

In retrospect, it is hard to say whether the romantics won or whether they were simply co-opted. Though the energy concept grew to encompass a wide range of natural phenomena, truth, beauty, and wisdom remained beyond its scope. With the introduction of the atom, nearly all forms of energy came to be understood in mechanical terms. But before we take this step, we must investigate one of energy's most mysterious disguises, potential energy.

CAN WE GET OUR MONEY BACK?

All the transformations of energy we have considered so far involve some noticeable change in the world; gasoline is burned, brakes heat up, and so forth. But with certain forces, especially gravity, something mysterious happens. A falling stone picks up energy as it drops. Where does this energy come from? Nothing has apparently changed, except the location of the stone, but energy has miraculously appeared.

The solution to this problem can be illustrated by using the example of a stone lifted by a hoist, as shown in Fig. 5-1. From the point of view of energy, this is the reverse of free fall, for energy is expended with no visible result other than a change in the position of the stone.

Again, from the newtonian point of view nothing very interesting is happening. Through most of its rise the stone is moving at constant speed, its weight exactly balanced by the force exerted by the rope. To set it in motion there had to be a brief instant when the force exceeded its weight, but from then on the motion is not accelerated. In terms of Newton's laws, the stone might as well be standing still. Yet from the

Except for a brief start up the forces are in balance

Afterward the stone rises at constant speed because the net force is 0

mg + "a little bit"

Thus, as the stone rises, there is no acceleration; yet work is being done by the man at the crank

↑mg
↓mg

But if the stone is released by cutting the rope at the top...

...it reaches the ground with the same speed it would have acquired had the work been done in the absence of gravity

Figure 5-1

point of view of the person pulling on the crank, there is a considerable difference. He is really working up a sweat. Were the stone standing still, he could set a brake on the hoist and go away; but no such simple process will raise the stone. He is putting work into the process, exerting a force on an object that moves. The work is the product of the weight of the stone mg and the distance h he raises it. Yet this work does not go into increasing the stone's kinetic energy; the counterbalancing force of gravity takes it right out again. Is that work lost forever?

The answer, of course, is no. If he cuts the rope, the stone will fall. Gravity will then work on the stone, and it will reach the ground with a kinetic energy equal to the work done by gravity. Since the distance moved was the same as that covered in the hoisting process and the force again is the weight of the stone, he has gotten back exactly the amount of work he put in. The stone reaches the ground moving at the same speed it would have acquired if he had done the same amount of work without gravity opposing his efforts.

Thus, gravity seems to be an "honest" force; work done overcoming its effect, even if that work produces no immediate reward in the form of motion, can be recovered later. Not all forces have this nice property; the work done dragging a stone across rough ground is forever lost.

Energy "stored" in this form is called *potential energy*. The sense of the word is self-explanatory. By raising the stone we have created a situation which has the potential of creating motion. Allowing the stone to return to its starting point will convert that potential to an actual motion.

The process of converting the potential energy into kinetic energy is a gradual one. When the stone has fallen only one-tenth of the way to the ground, gravity has done only one-tenth of the work it will finally do; one-tenth of the energy has become kinetic, the other nine-tenths remain potential. As the stone continues its fall, the potential energy is gradually used up and the kinetic energy increases. When the stone reaches the ground, the potential energy has all been converted and the work originally put into raising the stone appears as kinetic energy.

These relationships can be expressed mathematically in the form

Kinetic energy + remaining potential energy = work done in hoisting

or, in convential symbols,

$$\tfrac{1}{2}mv^2 + mgh = mgH \tag{5-4}$$

where h is the height of the falling stone, and H was its height when it started its fall. With the aid of this formula, for any height h we can calculate the speed of the stone, for its kinetic energy is the difference between mgh and mgH.

But Galileo could have done the same thing without introducing all these new concepts. If Equation (5-4) were applicable only to a falling stone, it would hardly be worth the trouble to write it down. But consider the roller coaster depicted in Fig. 5-2. The formula applies equally well to it. Once it is hoisted to the top of the first rise and released, it moves subject to only two forces: gravity and the support provided it by the rails. But the latter is perpendicular to the motion, and therefore does no work. The speed acquired in dropping a certain vertical distance is the same as for a freely falling body.

Here we see a more complex process. Just as the roller coaster draws on its supply of potential energy as it falls, it "puts it back" as it rises, slowing down but increasing its potential energy. Equation (5-4) tells us that anywhere on the roller coaster's track, its speed depends solely on how high it is above the ground.

Of course any real roller coaster loses some energy in the form of heat, through air resistance and friction in its wheel bearings. Thus it actually runs slightly slower each time it returns to any particular height.

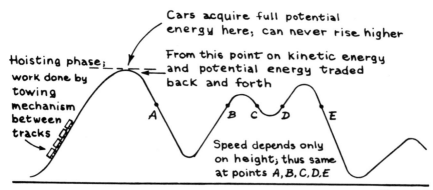

Figure 5-2

Gravitational potential energy has been a major energy source for the human race since before the dawn of civilization. From the primitive water wheel to the turbogenerators at the Grand Coulee Dam, we have exploited the potential energy of water as it descends to the sea. This is a self-renewing energy resource, because the ultimate source of water power is solar energy. Sunlight is absorbed in lakes and seas and converted into heat. This heat evaporates water, which is carried to the high clouds, to come down again as rain and keep the rivers flowing.

You may well be wondering at this stage *where* the energy goes while it is in the invisible "bank" of potential energy. If we stick to the strict action-at-a-distance theory of gravity, the answer is "nowhere," and the mystery remains. But once we get to the concept of field, we will find that the question does have a sensible answer.

We will also soon see that there are other forces that can store potential energy.

ENERGY AND ATOMS

When one looks below the surface of Equation (5-4), the law of energy conservation once again achieves a sort of unity. One of the greatest accomplishments of later nineteenth-century physics was the discovery that if matter is regarded as being composed of atoms, all the laws of thermodynamics can be understood by assuming that heat is nothing more or less than the energy of motion of these atoms. Imagine, as in Figure 5-3, a collision between two balls of soft clay of equal mass, heading toward each other at equal speed; after the collision they will be stuck together and standing still. Without even bothering to look into the internal structure of the clay balls, we find that total momentum is conserved, for momentum conservation holds without exception in mechanics, regardless of the level at which a system is observed. But the

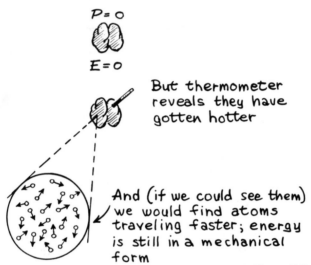

Identical clay balls of equal speed

$$P = mv - mv = 0$$
$$E = \tfrac{1}{2}mv^2 + \tfrac{1}{2}mv^2$$
$$= mv^2$$

After collision, momentum conserved but energy lost

$$P = 0$$

$$E = 0$$

But thermometer reveals they have gotten hotter

And (if we could see them) we would find atoms traveling faster; energy is still in a mechanical form

Figure 5-3

energy of motion is lost. Yet if we examine things on a deeper level, we find that the energy of motion has merely been transferred into motion of the atoms of which the clay balls are composed; this can be observed by noting the increase in temperature of the clay balls, for temperature is nothing more than a measure of the average energy of these atoms, as will be seen in Chap. 13. This motion is complex, chaotic, and random, and there is no way to reverse the process; to get all the atoms of a clay ball moving in one direction to restore the original energy of gross motion would be impossible. Momentum conservation also holds for this random motion, but there is no need to go into such details to establish that the total momentum of the clay balls has been conserved.

When we go from the macroscopic world of the clay balls to the microscopic one of the atoms of which the balls are composed, the law of conservation of energy regains its mechanical character to some extent. Not only does heat yield to this analysis; chemical energy can be seen as potential energy of the forces that bind atoms together. This was the source of the power of the concept to unify physics, for it led to a hope that all natural phenomena might ultimately yield to a mechanical interpretation if only their microscopic details could be understood. Then kinetic and potential would remain the only true forms of energy.

The example of the clay balls points up another advantage of the concept of energy. From the point of view of total momentum alone, there is no distinction made between the clay balls before and after the collision or between an elastic collision and one with clay balls. Energy does provide such a distinction—it provides a measure of motion that is independent of its directional properties. Indeed, the laws of elastic collision are derived from assuming that momentum and energy are simultaneously conserved. Applied together, the concepts of energy and momentum permit a more complete description of moving objects than either concept can by itself.

We can classify the three collision examples in Chap. 2 in terms of energy. In the first example, where the balls stick together, part of the kinetic energy is converted into heat. In the elastic collision, the kinetic energy is retained. The third example requires that a substantial amount of energy be converted into kinetic form at the moment of collision, perhaps by attaching an explosive cartridge to one of the balls.

BINDING ENERGY

The potential energy concept is a very useful tool for describing situations in which objects are bound together by forces of attraction. The earth and moon are bound in this fashion, as the earth and its sister planets are bound to the sun. In the microworld of the atom, the electrons are bound to their atoms by electrical attraction.

From the energy viewpoint, such objects are bound because they do not have enough energy to get away from one another. To move the moon away from the earth, one would have to put in energy in the form of work against the force of attraction.

For example, if an object leaves the earth's surface with a speed greater than about 11 km/s, it is free to "escape" the earth's gravity. It still slows down as it moves out, but gravity can never do enough work to bring it to a halt. No matter how far it goes, there will still be some energy left. This is the meaning of the term *escape velocity* heard so often since sending objects out into space became a commonplace. Escape would not be possible were it not for the fact that the force of gravity drops off as the square of the distance. Were the force to remain constant, for example, one could do sufficient work to stop the object and bring it back simply by going far enough. The formulas used earlier in the chapter, in which the dropoff of gravity with distance is ignored, are useful only on a terrestrial scale of distances, where vertical travel is small compared with the earth's radius.

In situations like this, it becomes convenient and logical to choose a zero point for potential energy in such a way that an object just free to escape has zero total energy. When this convention is adopted, all potential energies become negative. An object *gains* kinetic energy as it moves inward in response to an attractive force. Thus, its potential energy must decrease.

This becomes a convenient scheme for classifying the motion of objects under attractive forces. If the total energy is positive, i.e., the positive kinetic energy is greater in magnitude than the negative potential energy, the object is free to escape. If the total energy is negative, it is bound. There is a distance at which the potential energy itself is equal to the negative total energy of the object. Since kinetic energy cannot be negative, it can never go beyond this point.

As an example, consider the lunar missions flown by the Apollo astronauts. They approached the moon from a great distance; with respect to its gravity, they had positive total energy. To become bound to the moon in a lunar orbit, they had to slow down by firing their rockets in reverse, "giving away" energy. Had they not done this, having come from a position where they were not bound by the moon's gravity, they would have left it. The maneuver was not undertaken without some trepidation; once they had thus obtained negative total energy with regard to the moon, they could not escape it without putting energy in. Had their rockets been unable to restart and boost them to a positive energy with regard to the moon, they would have been trapped forever.

STARS, PLANETS, AND LIFE

If the notion of defining potential energy in such a way that it is always negative seems unaesthetic, consider that it accurately represents the process of creating a bound pair of objects. Were we to try to build the solar system, starting with the earth and the sun at a remote distance from one another, we would have to take energy *out* of the system to put the earth in its close-in bound orbit. Otherwise it would be free to return to its original remote location. Since it is reasonable to regard objects so remote from one another that they exert no significant forces as having no energy due to interaction, when energy is taken out to bind them, it is reasonable to regard them as having negative energy. When we study relativity, we shall see that this negative energy takes on a more concrete manifestation, a defect in mass.

This analysis is the basis of one of the more popular theories of how stars and planetary systems are formed. One starts with an immense, diffuse cloud of cool gas. As it collapses toward its center under its own gravity, potential energy is converted into motion of its molecules, or heat. Eventually, at the center, the temperature becomes high enough to kindle thermonuclear reactions, and a star is born.

As for the planets, suppose the cloud initially has a very slow rotation. At great distances this involves very small speeds, but as the cloud collapses, it must turn more and more rapidly, just as a skater or dancer spins more rapidly as she pulls in her arms. By the time one reaches the comparatively small dimensions of a star, the spin may be too fast to allow it to hold together. Some of the material is spun off or perhaps never reaches the center. It coalesces in smaller centers further out which have too little potential energy to reach thermonuclear temperatures, so cooler planets are formed. The fact that all the planets in our solar system move in orbits in the same direction, and the sun itself rotates in this direction, leads credence to this speculation.

Of course, one might expect that a planet might be formed big enough to become a thermonuclear star in its own right. This proves in fact to be rather common. A substantial fraction of the nearer stars in the sky can be seen, through a powerful telescope, to consist of double stars bound in orbits. Even when one of the partners is too small to be seen by its own light, its effect on the motion of its senior partner can reveal its presence. In many of these cases, it is probably far more reasonable to regard the unseen partner as a planet.

Even after the human race accepted the blow to its pride of finding the earth no longer the center of the universe, there was still some hope that the solar system might be unique, perhaps the result of some cataclysmic cosmic accident. But today it seems far more likely that planets are the rule, rather than the exception. Among the myriad stars and galaxies of stars, there must be countless planets as hospitable to life as our own. It would be foolish in the extreme to let our human pride delude us into thinking that life is confined to our poor corner of the universe, or even that the mystery of consciousness is reserved to us alone.

CHAPTER SIX

Completing the Job: The Classical Physicists' "World View"

With Earth's first Clay They did the Last Man knead,
And there of the Last Harvest sow'd the Seed:
And the first Morning of Creation wrote
What the last Dawn of Reckoning shall read.

—FITZGERALD, THE RUBAIYAT OF OMAR KHAYYAM

As WAS SUGGESTED in the preceding chapter, the concept of energy conservation shows its greatest power to unify physics when dealing with matter on the atomic level. By the latter part of the nineteenth century, physicists had a most enticing prospect open before them: it seemed likely that the universe was composed solely of atoms in motion. If all the forces between atoms could be understood, then *all natural phenomena* could be explained in terms of Newton's laws. All forms of energy might then prove ultimately to be mechanical in nature. There might prove to be a very small number of *fundamental forces*, like gravity, which might well have force laws nearly as simple as those for gravity. The story of this search, which nearly succeeded and still goes on today in a modified form, is the topic of this chapter.

ELECTRICITY AND MAGNETISM

One such fundamental force is that of electricity. Electrical phenomena were studied extensively during the eighteenth century by physicists and a host of scientific amateurs, not the least of whom was Benjamin Franklin.

Franklin grew up in the era when Newton was the most revered figure in the English-speaking world and science the noblest calling to which a human being could aspire. Having risen from poverty to modest wealth by the age of 40, Franklin retired from active participation in his printing business with the intention of devoting the rest of his life to science. Just 2 years later, in 1751, he published his *Experiments and Observations on Electricity*. The worldwide fame of this work brought an end to his scientific career; he became too valuable to the American cause as a representative abroad to ever again enjoy the leisure to pursue scientific studies.

Though Franklin's work set the standard for electrical research for more than a generation, it was entirely nonmathematical; he had ended his schooling at age 10, unable to master simple arithmetic. But his example inspired a French engineer, Charles Coulomb, to put the study of electricity on a sound newtonian basis through the discovery of the law of electrical force in 1789.

Coulomb perfected a device called the *torsion balance*, shown in Fig. 6-1. With its aid, he was able to prove that electricity, like gravity, obeys an inverse-square law

$$F = \frac{q_1 q_2}{R^2}$$

The masses in Newton's gravitational law have merely been re-

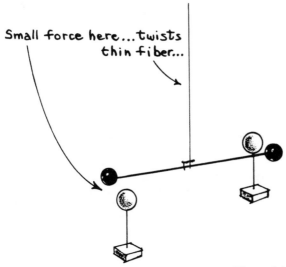

Small force here...twists
thin fiber...

Figure 6-1

placed by a new type of entity, *quantity of electricity* or *electrical charge*, denoted by the symbol q. Note that this law does *not* contain a constant like the g in Newton's law of gravitation; that is because, there being no previously agreed-upon units for amount of electricity q, one can *define* the unit of electrical charge so that the constant comes out 1. The unit of charge, called the *electrostatic unit* (esu), is defined as whatever charge must be placed on two objects 1 cm apart to produce a force between them of 1 g·cm/s².* Indeed, in Coulomb's time, there was no more reliable absolute measurement of the quantity of electricity than to measure the force produced.

Another distinction from gravity is that electrical force can be either attractive or repulsive. Electrical charge, unlike mass, can be either *positive* or *negative;* charges of *like* sign repel, while those of *opposite* sign attract. The formula indicates this fact, too: if the charges are of the same sign, their product is positive; if opposite, the product is negative.

Electrical forces are immensely more powerful than gravitational ones; the particles making up atoms are bound together by electrical forces 10^{40} times greater than the gravitational ones between them, and 10^{40} is a fantastically large number.† Though Coulomb's crude torsion balance was adequate to establish the nature of electrical forces, it took two further decades of patient development before the English experimenter Henry Cavendish, as mentioned in Chap. 4, was able to perfect a sufficiently refined torsion balance to measure the feeble gravitational attraction between a pair of lead weights. The purpose of the experiment, in Cavendish's view, was to establish the value of the constant g in Newton's gravitational formula, since in this case all the other quantities, the masses, distances, and force, were known. Since knowing the value of g permits one to calculate the mass of the earth from the known weight of an object on earth, Cavendish styled his experiment as "weighing the earth." (He might as well have said "weighing the sun," for from Kepler's description of planetary motion the accelerations of the planets are also known, and one can thus use the value of g to calculate the mass of the sun.)

---

* There are alternative sets of electrical units in which a constant *does* appear in Coulomb's law. Indeed, though all electrical units are based on the metric system, there are three separate sets, a situation that often provokes physicists to profanity when dealing with electrical computations.

† A convenient way of expressing very large or very small numbers is *scientific notation,* in which a number is expressed in a reasonably sized form and multiplied by 10 raised to some power. Thus, 10^{40} designates a 1 followed by 40 zeroes. Other examples of this notation are

$$524 = 5.24 \times 10^2$$
$$10{,}746{,}000 = 1.0746 \times 10^7$$
$$0.00000000000000029 = 2.9 \times 10^{-16}$$

When we reach the physics of the atom, where our ordinary scale of measurement is inconvenient, this notation will be indispensable.

The study of electrical forces was by no means completed with Coulomb's law. As the nineteenth century dawned, Count Volta, an Italian, developed the electrical battery, permitting for the first time a current, or steady flow, of electric charge. This opened a wide range of new experimental opportunities and brought researchers flocking to the field. Then, in 1820, the connection between electricity and magnetism was discovered, a connection that is nearly impossible to describe without the field concept. It is time we met Michael Faraday, regarded by many physicists as the greatest experimenter in the history of their science.

FARADAY AND THE CONCEPT OF FIELD

In London in 1812, a 21-year-old apprentice bookbinder presented himself to Humphry Davy, who had advertised for an assistant in his chemical researches. Michael Faraday's credentials consisted solely of a set of bound notes on Davy's public lectures. That these proved adequate was due in part to the ideology of the institution of which Davy served as director.

The Royal Institution had been founded by Count Rumford for the express purpose of improving the lot of the British working class through science. It was to provide a home for research to enhance their standard of living and serve as a beacon of light to encourage self-help through education. The London middle class found charities of this sort less costly than decent schools and a living wage. As the living embodiment of these ideals, Faraday could hardly be dismissed out of hand.

It soon became obvious that Faraday was a talented researcher in his own right. He gradually won his independence from Davy and, at the age of 34, succeeded him as director. Shortly thereafter, he abandoned chemistry and turned his hand to electrical research.

Though self-educated, Faraday was by no means unsophisticated. He was extremely well read in all branches of natural philosophy and rejected the strict newtonianism then prevalent. He was much influenced by the Jesuit Rudjer Boscovich, a native of Ragusa (now Dubrovnik) on the Dalmatian coast.

Boscovich, a contemporary of Franklin's, had argued that there was no need for separate concepts of force and matter. The ultimate atoms of matter could well be nothing more than points that served as centers of force. This idea is central to much modern speculation on the nature of matter. Faraday carried this viewpoint one step further. If force is to be the ultimate reality, he was sure that it must be based on something more substantial than action at a distance.

Faraday regarded the space between objects that exert forces on one another as being filled with something he called a *field*, which serves to transmit the force. As an aid to visualizing the field, he developed a pictorial scheme called *lines of force*, as shown in Fig. 6-2. These lines

Faraday and his wife. (The Royal Institution.)

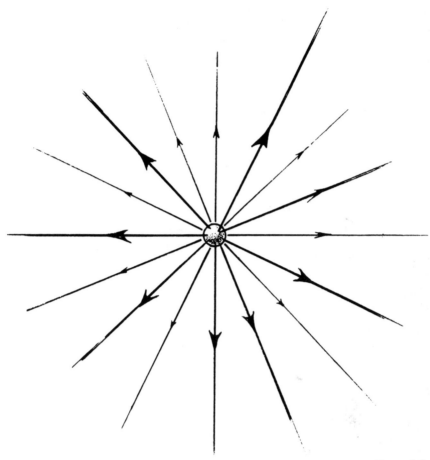

Figure 6-2

represent the force in two ways: the *direction* of the force at any point in space is along the lines, and the *strength* of the force is greatest where the lines are closely spaced. The absolute number of imaginary lines is of course arbitrary, but the relative number in various places on the graph is important. When one charge is regarded as fixed, the force any other charge introduced in the region can be deduced from the pattern of lines of force.

The power of this graphical device is not very evident when dealing with a single charge. In Fig. 6-3, the lines of force produced by a system of two opposite charges is depicted, alongside a sketch indicating how the force on a third charge at one particular point is deduced by the more conventional method of vector addition. In this situation, one of the rules for drawing these lines of force is that each one begins on one charge and ends on the other. When even more complex arrangements

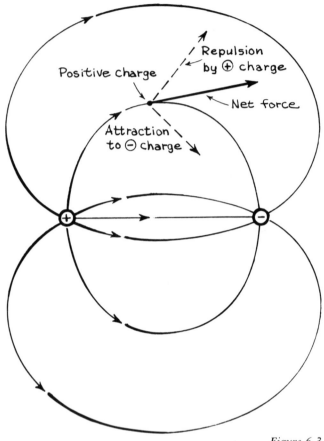

Positive charge

Repulsion
by ⊕ charge

Net force

Attraction
to ⊖ charge

Figure 6-3

are considered, it is obvious that the value of the lines of force as a means of visualization is very great. They also provide the natural explanation of the inverse-square law, discussed at the end of Chap. 4.

The power of the field concept becomes more obvious when we move from electricity to magnetism, a complicated force that is not a simple matter of attraction and repulsion. First of all, magnetic fields are generated only by *moving* charges; if the charges are standing still, there is no magnetic field. The simplest case is the magnetic field of a steady electric current, shown in Fig. 6-4. The lines of force do not radiate out from a current-carrying wire but form rings around it.

The complications do not end there. A magnetic field exerts a force only on *another* moving charge. And the direction of the force is not along the field line but at right angles to both the line of force and the direction of motion. Thus a charge moving toward the wire will be deflected along the length of the wire; one moving parallel to the wire will be attracted to it or repelled from it, depending on the sign of the charge. Obviously, it would be hard to visualize these geometric complications on the basis of a formula alone or come up with a natural explanation of them in terms of action at a distance.

Still, it must be emphasized that the two points of view, field and action at a distance, are completely equivalent *as long as the field is constant in time*. Thus, Faraday's approach was not taken very seriously until his work was refined and extended by a very sophisticated mathematical physicist, James Clerk Maxwell.

Examining Faraday's discoveries concerning the interrelations of electrical and magnetic fields in order to put them on a sound mathematical basis, Maxwell discovered a startling implication: if the charges generating a field move or disappear, the effects of this change will *not* be communicated instantaneously to a remote charge. The change in the field instead moves out at a high but finite velocity, the velocity of *light*.

That this discovery makes it essential to assign some reality to the field itself is very easy to show. Consider two charges interacting, as depicted in Fig. 6-5. Now let us displace charge *A* to the right. Since *A* is responding to the field of *B*, the force on it will continue to point to *B*. But *B* will be, for a brief period, "unaware" that *A* has moved; the force on *B* will continue to point to where *A* was!

Thus, we find the electrical force violating, at least temporarily, Newton's third law; the forces the two charged bodies are mutually exerting on one another are *not* opposite. As a further disastrous consequence, if the two charges are free to move, momentum is not conserved, since if the forces are not opposite, there is a net increase in momentum to the right of the line joining the charges.

Must we therefore give up Newton's laws and momentum conservation when dealing with electric and magnetic fields? Not at all, for Maxwell

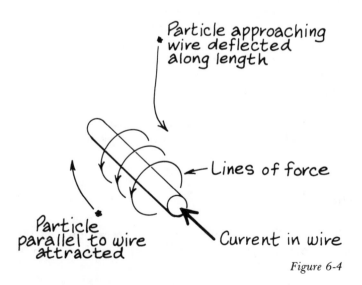

Figure 6-4

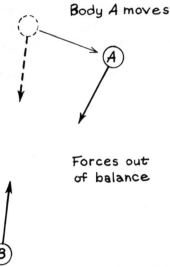

Body A moves

Forces out
of balance

Figure 6-5

and his successors were able to show that when an electric charge is moved, an additional force is required above and beyond that needed to accelerate an uncharged mass. This force is the consequence of the interaction of the charge *with its own field,* regardless of whether or not other charges are present. In the process, forces are exerted on, and thus momentum is imparted to and work is done on, the field itself. This momentum and energy are transported at the velocity of light and may be recovered elsewhere by other charges. Figure 6-5 shows a violation of Newton's laws only if we consider space as containing only bodies A and B; the field is there, too, and it carries energy and momentum and exerts quite tangible forces on accelerated charges. If the interaction of each body with the field is considered in detail, it is found that Newton's laws, and also momentum and energy conservation, are preserved.

Another useful feature of the field concept is that it lends more reality to the concept of potential energy, which we regarded as a mathematical fiction in the context of the preceding chapter. Maxwell showed that whenever a field exists, there is energy distributed throughout the region of space where it is present. It is this bank of energy on which an object draws when it converts potential energy into its kinetic form.

AN ALMOST FINAL UNITY

The capstone of Maxwell's achievement was the recognition that the speed of propagation of a changing electromagnetic field is the velocity of light. This Maxwell was able to deduce from the relative strength of electrical and magnetic forces, as measured by Faraday. At this point it seemed logical that light itself must be one such form of electromagnetic disturbance, produced by a charge in a regular, repetitive, accelerated motion so that a repeated *wave* of energy and momentum is generated. (This will be discussed further in Chap. 7.)

Maxwell's achievements, unifying electricity, magnetism, and light, were the high-water mark of classical physics, its last great triumph. His principal work, the *Treatise on Electricity and Magnetism* published in 1873, can be compared to Newton's *Principia* in significance. Like the *Principia*, Maxwell's *Treatise* is a work of synthesis, combining the efforts and thoughts of his predecessors and a few new ideas to form a complete and all-encompassing theory. Like Newton, Maxwell worked best in solitude, preferring his family's isolated estate in Scotland to university life.

Let us now examine where Maxwell's achievements left physics, for in a sense they mark a breaking point in the story that began with Galileo.

A physicist in the latter years of the nineteenth century could believe there was nothing in the universe but matter and a mysterious substance called electricity. Both of these generated *fields*, electrical and gravitational, which had a reality of a different order than that of matter. The fields produced forces which acted on matter in accordance with Newton's laws. The properties of matter derived from those of the atoms of which it was apparently composed, and once these were catalogued, all natural phenomena would be completely understood.

There seemed to be only two remaining tasks, each very well defined and even well on the way to completion, for subsequent generations of physicists. One was to complete the study of atoms, and the other was to cap the unity of physics with a purely mechanical explanation of the electromagnetic field. Maxwell believed, in fact, that he had actually achieved this explanation in the theory of the *aether,* which will be dealt with in Chap. 8. Many of the finest scientific thinkers of the time were advising young men not to go into physics, as nature held no further secrets of importance in that realm. Man could at last dream of a final understanding of nature on its most fundamental level. It is a curious fact, and perhaps no coincidence, that this sentiment came during the late Victorian era, one of the most complacent times in modern European history, when the middle and upper classes in Europe felt they had created a nearly perfect society, likely to endure forever, with merely a few minor social problems remaining for reformers to clean up.

But, along with the "old order" that had supported it so well under royal patronage, with the material assistance of the bourgeoisie, classical physics was headed for a smashup. The study of atoms led to inquiries into their structure, which proved full of paradoxes unresolvable within the newtonian framework, necessitating the development of the quantum theory. And Maxwell's aether theory failed spectacularly in its most significant experimental test, a failure that was to force a reevaluation of such elementary concepts as space and time, leading to relativity. This is the story of the remainder of this book.

EPILOGUE: THE NIGHTMARE OF DETERMINISM

At several points in the exposition of classical mechanics, we have commented on how the thinking of physicists often shows signs of the influence of the spirit of the time in which they lived. That their ideas in turn influenced this spirit may seem surprising, in view of the abstract nature of physics and its near isolation from other fields outside the natural sciences. But through the medium of philosophy, physical thought does from time to time filter down into the general intellectual currents. It seems fitting to close the story of classical mechanics with some remarks about its impact on social thought.

The spectacular rise of physics in the two centuries leading up to the Victorian era led thinkers throughout the intelligentsia to give considerable credit to its claims of universality. Given the classical physicist's "world view," it is reasonable to believe that everything that happens in the universe is no more than a manifestation of the motion and interaction of the constituent atoms of matter. This motion is governed by perfectly deterministic laws; the mathematical physicist Laplace speculated that if one could only observe at some instant every atom in the universe and record its motion, both the future and the past would hold no secrets. Put another way, all of history was determined, down to the last detail, when the universe was set in motion. The rise and fall of empires, indeed, the heartbreak of every forgotten love affair, represent no more than the inevitable workings of the laws of physics; the universe marches on like a gigantic clockwork.

What room, then, for free will, for salvation and damnation, for love and hate, when the most trifling decision any human being can make was determined 10 billion years ago? It gave the ethical thinkers of the nineteenth century something to ponder. Admittedly, it is inconceivable that one could actually achieve the omnisicience speculated on by Laplace. But the fact that it was possible in principle was viewed by materialist-oriented thinkers as a genuine nightmare.

Later we shall find that modern physics has at least partially resolved the dilemma by introducing an element of chance into the heart of the infernal clockwork and by making the distinction between what can be known in principle and what is actually known by observation a crucial one at the core of the theory; thus, we shall return to this point again in the context of quantum mechanics.

But even those thinkers untroubled by this admittedly abstract nightmare felt the impact of physical thought. For the first time, all the details of a tremendous range of natural phenomena were understood in terms of a remarkably small number of principles. Classical physics became the model of what human knowledge should be. Many of the social thinkers of the nineteenth century tried to emulate this universality and precision, seeking general laws to explain history and human behavior. Consider, for example, the writings of Karl Marx. Consider also the efforts of Freud to explain such aspects of society as religion and mythology in terms of his picture of the human mind.

This sort of *social determinism* is particularly reflected in the tactics of the celebrated defense attorney Clarence Darrow. Defending a client who was patently guilty of the offense as charged, Darrow would point to him as the prisoner of his own heredity, placed in an environment not of his own choosing. Under such circumstances, following from a chain of causes leading back to time immemorial, what was the meaning of "responsibility for one's actions?" And how could society presume to punish a man for a situation it had itself created, in which the victim was as powerless to modify his fate as the hands of a clock are to refuse to turn?

But, as we shall see, the twentieth century has taught physicists another lesson: as one moves from one level of reality to another in the study of nature, the laws and concepts used to describe it themselves must change. Though the laws describing the behavior of atoms are, at least in principle, the basis for the behavior of larger objects, it is inconvenient and perhaps even impossible to so use them in practice. On the basis of this, a physicist might well suspect that, for example, even if psychology were to become a perfectly exact science, it would be of little value in understanding society. Most problems in human knowledge must be solved on their own "natural" level, and to look to a "deeper" level may be instructive in itself but rarely helps to solve the problem that started the inquiry.

CHAPTER SEVEN

Waves

WITH THIS CHAPTER we finally drop the study of mechanics and devote a brief interlude to the motion of waves. One might well imagine that as long as the word *motion* is used, the subject is still mechanics, but the wave is a peculiar concept that appears in many guises. It is particularly ubiquitous in twentieth-century physics, so an understanding of the language of waves is essential to the remainder of this book.

Wave motion is not a mechanical phenomenon because a wave is not a material *object* but a *form*. It cannot be assigned a mass, and the concept of acceleration is absolutely valueless for dealing with waves. The motion of a wave is vastly different from that undergone by the medium in which it travels; in fact, we can have a wave without any movement of matter at all. The wave follows its own laws, regardless of the underlying physics.

What, then, is a wave? It is a pattern, a form that moves. It can be a deformation of a material object, as in the case of a music string or waves on the surface of a body of water. It can also be a pattern in a field, such as light or radio waves. And these examples by no means come close to exhausting the roll of wave phenomena.

MOVING BUMPS

In this chapter we shall start with waves in their simplest form, single bumps or *wave pulses* traveling on a one-dimensional object, such as a string. We shall then move to continuous or repeated waves, and next to waves in two or three dimensions, building ultimately to the explanation of the experiment by which the wave nature of light was finally convincingly demonstrated.

There is no point to going any farther without introducing concrete examples, so we shall begin with the simplest, the single wave pulse. The easiest example to visualize is that of a bump on a taut string. Left to itself, such a string would remain straight. If some outside agent deforms the string, the tensions will pull the bump down, as illustrated in Fig . 7-1. But Newton's third law remains operative here; if the portion of the string to the right of the bump pulls it down, it must in turn be pulled up. By the time the original disturbance has been eliminated, a new one has been created to its right; this process continues, and the result is a bump that travels to the right. Note that it is the bump that travels, not the string, which experiences an outward deflection, followed by a return motion, actually perpendicular to the motion of the wave itself.

There is no need to dwell on the mechanical details of the process. True, a real motion of the string gives rise to a very different apparent motion of a bump. But we shall see that it is perfectly possible to understand the latter motion without troubling ourselves with the details of the former. The motion of the medium can almost be ignored in the study of waves.

To make this point clearer, let us choose another example totally devoid of any connection with the laws of mechanics. Imagine a marching band formed up in a long single line. Each member of the band is given the instruction, "Watch the player on your right and do what he does one beat of the music later." We then go to the end of the line and ask the musician there to take two steps out, then two steps back.

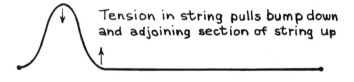

Tension in string pulls bump down and adjoining section of string up

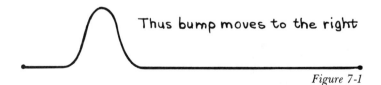

Thus bump moves to the right

Figure 7-1

The resulting effect is illustrated in Fig. 7-2; viewed from above, a "bump" travels down the line of musicians from left to right, yet not one of them has moved either to *his* left or right. And this is by no means a mere analogy, but an actual wave phenomenon in the full meaning of the word, except that the "medium" is not continuous, as it is in most of the cases we will study. Nonetheless, wave laws can be applied in exactly the same way they are to more physical waves.

This gives us the opportunity to cite one important property of waves: they travel at a constant speed that depends on the nature of the medium, not that of the wave. The wave travels a distance of one interval between musicians for each beat of the music. This is because of the marching instructions, not the form of the wave. Had we asked the end player to take *three* steps out and back, a larger wave would have been formed, but it would still have traveled the same speed. In fact, the elementary concept in wave motion is that of a wave that travels at constant speed without changing its form. Any other kind of wave behavior must be expressed in these terms, by methods that will be shown. Thus there is no need to prove that this form of behavior holds true for the wave on the string. If it does not, it merely shows that the string is a poor example. As it happens, a tightly stretched string is an

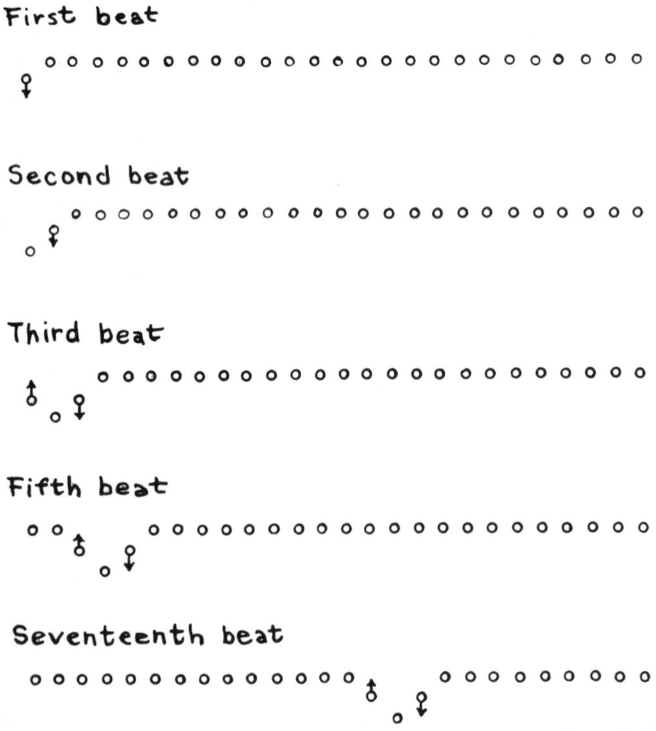

Figure 7-2

Before meeting

When meeting

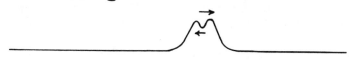

Afterward

Figure 7-3

excellent wave medium. If the tension is high or the string light, the acceleration of the string in returning from a distorted condition is great and the wave moves fast. If the string is heavy or the tension weak, the wave moves more slowly.

SUPERPOSITION

There is one universal law that has a crucial role in the study of wave motion, however, a role comparable to that of Newton's laws in mechanics. This law is called the *principle of superposition*. It is similar in spirit to the law of the same name in mechanics but should not be confused with it. It rests on the fact that the presence of one wave does not alter the ability of the medium to transmit another wave. Thus, two waves can pass through one another on a medium without changing their form. A simple example for the case of opposite-traveling waves on a string is illustrated in Fig. 7-3. The small wave merely appears as a bump on the big one as they pass.

Stated in quantitative terms, the principle of superposition ensures that *the displacement produced by two waves at the same point is merely the sum of the displacements produced by each alone.*

The results of this principle are most interesting when applied to waves of equal size. Figure 7-4 illustrates the application of the principle of superposition to two identical waves, first displaced in the same direction and then oppositely displaced. The second case, Fig. 7-4*b*, is perhaps the most amusing. For a brief instant, the string is absolutely flat; but since the segment just left of the center was just an instant previously moving down and that to the right moving up, this motion will persist and the waves will be recreated. A little reflection will show that there is one point on the string, halfway between the waves as they approach, that will never move at all.

The simultaneous presence of two waves at a single place is called *interference.* When the waves both act in the same direction, as in Fig.

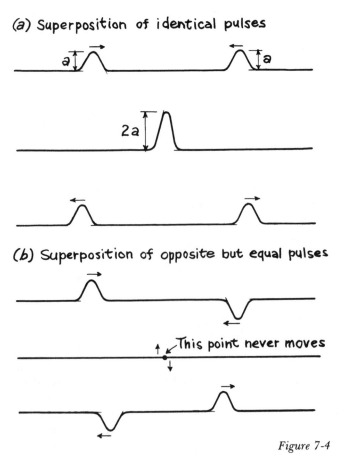

(a) Superposition of identical pulses

(b) Superposition of opposite but equal pulses

This point never moves

Figure 7-4

7-4a, the interference is called *constructive*. The phenomenon is Fig. 7-4b is known as *destructive interference*.

The only connection between the mechanical and wave-superposition laws is that when the origin of the wave is a mechanical process, as in waves on a string, the wave law can be proved as a consequence of the mechanical one. But many wave phenomena are nonmechanical, so it is best to regard the wave rule as standing independently.

The most interesting applications of the principle of superposition are the "backward" ones. These are the cases in which we analyze a wave and predict its future development by imagining it as the sum of two other waves, similar in spirit to the argument by which Galileo analyzed projectile motion, where he decomposed a complex curved motion into two simple ones.

As an example, let us ask what happens if we form a bump in the *middle* of a string. It is equally free to move in both directions, with no inherent tendency to go one way or the other. What will it do?

The question is easily answered if we take note of the fact that this situation is exactly that which exists for the instant the two waves in Fig, 7-4a are exactly together. There is no difference in the appearance or motion of the string in the two situations. Even though in one case the bump formed as the fusion of two waves and in the other we formed it artificially, there can be no difference in the subsequent behavior of the wave. We thus predict that the bump will split into two waves, each of half its height, moving off in opposite directions, and observation verifies that this is the case.

REPEATED WAVES

The most important wave phenomena concern not individual wave pulses but trains of repeated, identical waves. These follow no laws different from those of individual wave pulses, so all we need add is a terminology to describe them. Figure 7-5 illustrates this terminology. The *wavelength*, for which the lowercase Greek letter lambda (λ) has become the conventional symbol, is the interval at which the pattern repeats. The *amplitude* measures the size of the displacement produced by the wave. One more word is required to complete the description: since the wave is moving, any point on the medium goes through a motion that repeats itself as each wavelength passes. The number of times per second this happens is called the *frequency*, usually denoted by the Greek letter nu (ν). Frequency is measured in units of *cycles per second* renamed *hertz* (Hz) in honor of the discoverer of radio waves. Some-times, in place of wavelength, it is more convenient to use the *wave*

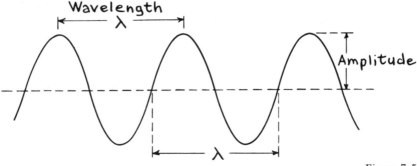

Figure 7-5

number k, the number of waves per meter. This is simply the inverse of the wavelength; that is, $k = 1/\lambda$.

Of course, the wavelength and frequency are not unrelated, since the wave does travel at a fixed speed. For example, if a wave has a frequency of 5 Hz, i.e., if 5 complete cycles occur each second, and if each wave is 4 m long, the wave must be traveling 20 m/s. This relationship can be summarized in the formula

$$c = \lambda\nu \qquad\qquad (7\text{-}1)$$

where c is the (also conventional) symbol for the wave velocity. This is not a physical law in the usual sense, but merely a relation that follows from the definitions of wavelength and frequency.

The smooth wave shown in Fig. 7-5 is known as a *sine wave* because the mathematical description of it uses the trigonometric sine function. A sound wave that has this shape is heard as a pure musical tone whose pitch is determined by the frequency. A sine light wave gives a pure spectrum color. Waves, however, can have almost any imaginable shape. As long as the shape is faithfully repeated for many wavelengths, the principle of superposition can be used to build them up as combinations of sine waves of different wavelength and amplitude.

When a wave is confined between fixed boundaries, such as the ends of a music string, we get the beautiful pattern known as a *standing wave*. One such pattern is shown in Fig. 7-6. All musical instruments generate their sounds in some such fashion.

A standing wave can exist on a string only if it fits in such a way that the ends do not have to move. Since a wave pattern crosses the centerline once each half wavelength, the only waves that survive are those with half wavelengths exactly 1, $\frac{1}{2}$, $\frac{1}{3}$, . . . times the length of the string. The case of 3 half wavelengths is shown in Fig. 7-6. The points that do not move are called the *nodes* of the wave.

Since shorter wavelengths mean higher frequencies, the pattern with 1 half wavelength, called the *fundamental,* has the lowest pitch. The shorter waves are called *harmonics* or *overtones.* The principle of super-position allows many harmonics to be present at the same time. This is what gives musical instruments their character, since a pure sine wave has a boring, rather mechanical sound.

A radio transmitter sets up a standing electrical wave in its antenna. A laser generates a standing light wave, confined between mirrors at either end, with a mechanism for pumping energy into the wave. When we get to the quantum theory, we will meet up with a very interesting type of standing wave.

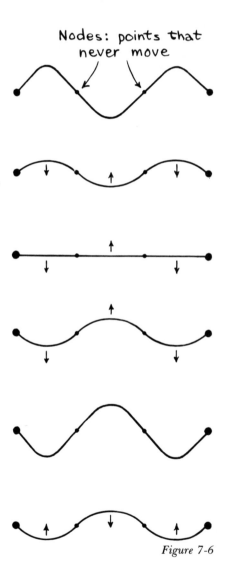

Figure 7-6

WATER WAVES AND SUCH

So far we have considered only one-dimensional waves, i.e., waves traveling on a string, a linear medium. Even more familiar are two-dimensional water waves—bumps in the form of long ridges on a two-dimensional surface of a body of water. The next step in the discussion of waves is to consider them in two or three dimensions. Again, no new physical principles are involved; the problem is simply one of description.

We will begin the treatment by reverting to the simple situation with which this chapter began—a single wave pulse. However, this time let it be a bump not on a string but on the surface of a pond. As the bump recedes under its own weight, adjoining regions are forced up by the increased pressure, beginning the propagation of a wave. A sectional view of the process is depicted in Fig. 7-7a, and the logic of the situation

(a) Sectional view

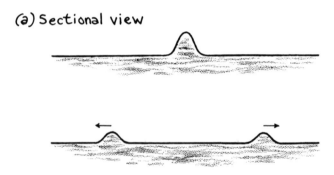

(b) View from above

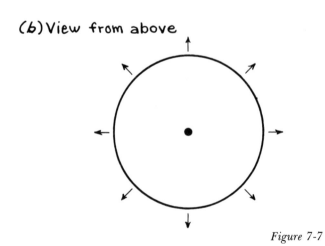

Figure 7-7

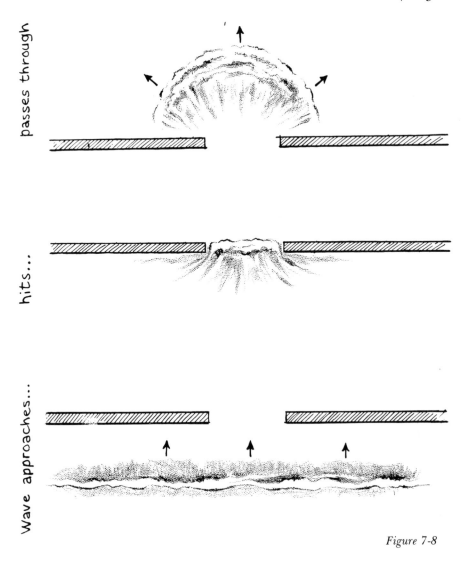

passes through

hits...

Wave approaches...

Figure 7-8

is exactly that used when we considered the effects of a bump in the center of a string. But this time the wave is free to move in all directions, not just two linear ones. Having no reason to prefer one direction over another, it must move out in all directions. The result is the familiar expanding circle of the ripple on a quiet pond, Fig. 7-7*b*.

Even more common to the ordinary experience is the case of continuous repeated *linear* waves, ones that take the form of long lines, such as the breakers rolling into a beach. One interesting property of waves of this type is the way they behave when they go through gaps in barriers, such as the opening in a breakwater. Figure 7-8 illustrates this case. If the opening is comparable to the length of the wave, each successive wave produces a "bump" in the opening, much like the source of the circular ripples of Fig. 7-7. The result is a circular wave pattern inside the breakwater. This phenomenon is known as *diffraction*. If the width of the opening is much larger than a wavelength, this will not happen; instead the waves will continue through as linear waves, with a slight fanning out at the ends.

Waves can also exist in three dimensions. The most familiar example is that of sound waves. The crest of a sound wave is a region in which the air molecules are packed closer together. The analogue of Fig. 7-7 for this case produces an expanding wave in the form of a *sphere;* similarly, just as *linear* water waves are observed, so are three-dimensional *plane* waves.

Both two- and three-dimensional waves exhibit a property called *refraction.* This is an effect that occurs when a wave passes a boundary from one medium to another in which it moves more slowly. The end of the wave inside the new medium is moving slower; therefore, since it remains connected to the faster-moving end in the old medium, it must have turned, becoming more nearly parallel with the boundary, as shown in Fig. 7-9. If the change in speed is gradual, so is the turn; this is why water waves nearly always strike a beach parallel to the water line, regardless of the direction they had in the open sea. As waves enter shallower water, they slow down, and the resulting turn brings them straight into the beach.

HOW LIGHT WAS SHOWN TO BE A WAVE

We now have the tools to examine a very significant interference effect, the one used by Thomas Young in 1789 to settle a long-standing controversy over the nature of light. This controversy dated from the time of Newton, who did extensive studies of optical effects, many of

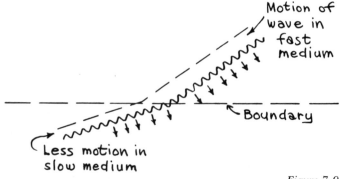

Figure 7-9

A two-slit interference pattern. (Courtesy of Brian J. Thompson The Institute of Optics, University of Rochester.)

which began in the same plague-enforced absence from Cambridge that led ultimately to the *Principia*. Newton favored a theory that represented light as a hail of tiny particles. Huygens favored a wave theory. Though the evidence available at the time to distinguish the two theories was skimpy, it must be admitted that even then Huygens' point of view seemed more plausible. But such was the prestige of Newton's name that few physicists dared finally discard the particle theory, despite mounting evidence for such effects as diffraction and constancy of speed, easy to account for in a wave theory but seemingly unnatural for particles.

Thomas Young, whose work represented the *coup de grâce* to Newton's point of view, was a man of truly protean talents. While still in his prime he resigned a physics professorship to carry on a medical practice, and a period during his later years was spent cracking the code of the Rosetta Stone, the key to the Egyptian hieroglyphics.

The experiment Young chose was that of *two-slit interference*. It is easiest to visualize in two dimensions, which brings us back to water waves and the hole-in-the-breakwater example of Fig. 7-8. Only this time, imagine a breakwater with *two* small gaps. The waves striking the breakwater will produce two synchronized circular patterns inside, and in due course these will overlap, as shown in Fig. 7-10.

At a point on the shore opposite and halfway between the two gaps, the waves always meet crest to crest because this point is equidistant from both gaps; the waves from each gap arrive simultaneously, and the

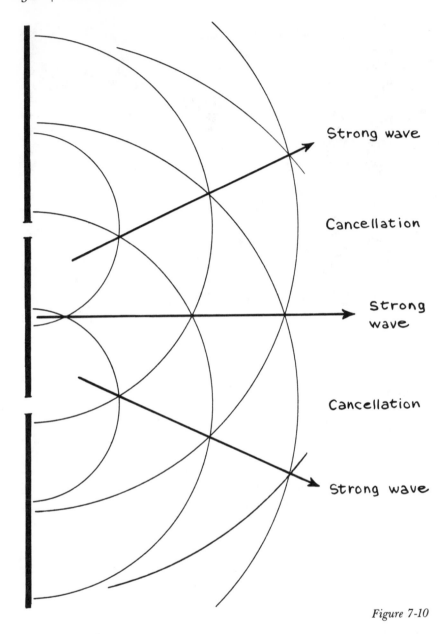

Strong wave

Cancellation

Strong wave

Cancellation

Strong wave

Figure 7-10

interference is constructive, giving a wave twice as high as that from one gap. As we move along the shore from this point, the synchronization is destroyed, for we are closer to one gap than to the other. Eventually we reach a point where the crests of the waves from one gap meet the troughs in the waves from the other. Here the interference is *destructive:* the waves are small or absent altogether. Moving further, we reach a point where the wave from the nearer gap meets the *preceding* wave from the farther gap. Once again the interference is constructive, and the waves are high. If we continue, we again reach a point of destructive interference, and so on. The rule is really quite simple: if the difference between the distance to one gap and the distance to the other is an integer (whole number) multiple of a wavelength, the interference is *constructive*. If it is $\frac{1}{2}$, $1\frac{1}{2}$, $2\frac{1}{2}$, $3\frac{1}{2}$, ... wavelengths, *destructive* interference results.

Now to Young's experiment: replace the breakwater with an opaque screen, the gaps with narrow slits. On a sheet of paper well back from the slits, one observes a pattern of bright and dark bars parallel to the slits—a bright one in the center, dark ones to either side, and so on. If the spacing of these lines is measured, a little work with geometry enables one to calculate the wavelength of the light. The result is fantastically small: light waves have wavelengths which range from 0.00007 cm for red light to 0.00004 cm for blue light. The slits must be very narrow and close and the viewing screen well back from the slits to make the effect visible.

WHAT KIND OF WAVE IS LIGHT?

Young's experiment convinced the last of the doubters that light must indeed be a wave phenomenon. But the question still remained: what sort of wave? It was Maxwell who, by perceiving the relation between electromagnetism and light, provided the answer.

The picture Maxwell gave is illustrated in Fig. 7-11. It rests on two facts, both discovered by Faraday: (1) whenever an electric field is changing, it generates a magnetic field, and (2) the magnetic field produced thereby is at right angles to the electric field. The relationship is reciprocal: a changing magnetic field creates an electric field at right angles to itself.

Without asking how such a peculiar combination of fields can be created in the first place, it is clear that the arrangement illustrated in Fig. 7-11 is self-perpetuating. The changing electric field generates a magnetic field which, since it too is changing, regenerates the electric field. The process continues indefinitely. The wave moves in a direction perpendicular to both fields.

Maxwell calculated the speed at which such a wave would travel and found it agreed exactly with the speed of light. The coincidence was too striking to be ignored. Maxwell concluded that light itself must be an electromagnetic wave.

How such a wave could be produced was evident from the laws of electricity and magnetism. Suppose an electric charge is going through some sort of repeated, periodic motion. Motion in a circle would do, as

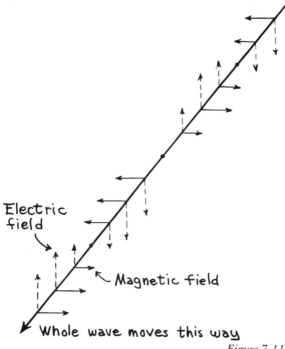

Electric field

Magnetic field

Whole wave moves this way

Figure 7-11

would simple back-and-forth oscillation. In either case, both an electric field and a magnetic field are generated, for the charge is in motion. The field is also a changing one, because the position of the charge is changing, and field strength varies with distance. Thus, any object which is giving off light must contain electric charges oscillating very rapidly, since the frequencies of light waves are in the range of 10^{14} Hz, or 10 trillion oscillations per second!

Since any object oscillating so fast would have to be very tiny, observing its oscillations directly seemed a hopeless task. But Maxwell reasoned that slower oscillations must also produce electromagnetic waves, whose strength would be sufficient to detect by electrical means. At frequencies of a few thousand hertz, comparable to those of sound waves, it is possible to set large quantities of charge in motion on electric wires. Twenty years after Maxwell published his theory, in 1887, the German physicist Heinrich Hertz was able to do the trick.

Hertz set up violent electrical oscillations in an electric circuit. In another circuit several feet away in the laboratory, not connected to any source of power, electrical oscillations of the same frequency were produced. The frequencies of the oscillations were easy to measure; from the interference properties of the waves, the wavelengths could be estimated. Thus, the speed could be computed, and it agreed with the speed of light. After these experiments, few could doubt that Maxwell had been right. Hertz's waves later became the basis of modern radio communications.

Shortly after the work of Hertz, x-rays were discovered. These proved to be electromagnetic waves, too, but of frequencies about a thousand times *greater* than those of visible light. Later, the gamma rays emitted by radioactive substances proved to be electromagnetic waves with frequencies a thousand times greater still!

Today, we have a technology for handling electromagnetic waves of nearly any desired frequency. Maxwell's theory of electromagnetism has proved probably the most striking example of a "pure" scientific theory leading to practical results that could never have been stumbled upon by accident.

CHAPTER EIGHT

Surprise: The Earth Doesn't Move!

S TRUCK AS THEY WERE by the beautiful description of a light wave given by Maxwell, it was nonetheless hard for the physicists of his time to imagine a wave without also imagining some sort of medium for it to travel in. Sound waves have air, ocean waves propagate on water, and so on. Furthermore, in the case of these material waves, their behavior could be attributed to the mechanical properties of the medium. The implied invitation was irresistible: find the medium, and the behavior of light (and thus ultimately of electrical forces) might thereby be shown to result merely from Newton's laws, putting the final capstone on the unity of physics. To this hypothetical medium Maxwell gave the name luminiferous (light-bearing) aether. Light would represent waves in the aether, and electrical and magnetic forces would be transmitted by aether pressures and flows.

This idea disturbed Faraday. The field, all by itself, was quite enough reality to suit him. But his line of thought was too far ahead of its time to appeal to his contemporaries, whose instincts were still entirely newtonian. Furthermore, Faraday genuinely admired Maxwell's mathematical gifts and thus hesitated to criticize the aether theory too forcefully. If it was to fall, it must be by experiment, and for that the time was not yet ripe.

It could easily be shown that this aether would have to be a rather peculiar substance indeed. From the fact that light propagates at the same speed throughout the universe, it is clear that it would have to fill space uniformly. In particular, this made it unlikely that it was affected much by gravity. And since the motions of the planets had by then been studied to high precision and observed over millennia without any change, it was clear that the aether must offer at most an infinitesimal resistance to the passage of ordinary matter. Indeed, the most likely possibility was that it could freely penetrate ordinary matter without exerting the least force. Only electrically charged objects could interact with the aether. And from the fantastic velocity of light, 3×10^8 m/s, it was clear that the aether must be low in density, yet very "hard," for both these properties contribute to a high speed for waves.

What a troublesome substance for the poor experimental physicist to get a handle on! Transparent, incompressible, and difficult to exert a force on—how could one hope to study it or even prove its existence, aside from the indirect evidence provided by light itself?

Nonetheless, a few opportunities presented themselves. The most promising was that any point on the earth must be moving through the aether at a considerable speed. Our planet clips along in its orbit at the rather brisk pace of 30 km/s; add to this a somewhat smaller speed resulting from the earth's rotation, and whatever comes from the motion of the sun, and you expect a considerable "wind" of aether "blowing" through the laboratory. It might be small compared to the 300,000 km/s of the velocity of light, but the aether wind should still be strong enough to have observable consequences in the behavior of light on earth.

The first experiment capable of detecting this aether wind was performed by a young professor at the Case School of Applied Sciences in Cleveland, Ohio, Albert A. Michelson, with the aid of Edward W. Morley, a chemist from Western Reserve University. The results were announced in 1887. That one of the most significant experiments in the history of physics could come from two relatively obscure schools in an industrial boom town on the shores of Lake Erie came as a bit of a surprise both in the United States and abroad. Though America had produced a few scientists of note by the end of the nineteenth century, it was still very much of a scientific backwater, especially in the more abstract areas of physics. The great centers of learning, the private universities of the East, were primarily concerned with educating America's governing elite; scientific studies rated low in their standard of values. Thus, American science was dominated by the technical schools and land-grant colleges, where it took on a very gadget-oriented, pragmatic bent that still survives to some extent today as an element of the American "style" in physics. Indeed, until the United States absorbed a major portion of the European scientific community as World War II refugees, young Americans bent on becoming first-class research physicists were well advised to go abroad to finish their education, especially if they were interested in fundamental problems.

Michelson was no exception to either trend. Graduated from Annapolis in 1873, he was assigned to work at the Nautical Almanac Office. Because of the traditional interest of all navies in optical and astronomical measurements, he had a great deal of freedom to pursue experiments in these areas. Bitten by the experimental "bug," he followed a well-established route, taking leave from his naval duties in 1879 for several years of polishing in Berlin, Heidelberg, and Paris. In Europe he found the problem of detecting the aether occupying the center of attention among those physicists who were experts on optical phenomena, and he made his first attempt at his celebrated experiment. Though he failed to find the aether wind, his first instrument was too crude to be sure the failure was not purely an instrumental error.

The trademark of this man, who was to become America's first Nobel laureate, was his ability to conceive instruments of high precision on a grandiose scale that transcended the traditional idea of a laboratory device. Michelson chose the aether-wind experiment at least partly because of the opportunity it provided to show his imaginative virtuosity at instrument design. The European experience also settled his career plans: shortly after returning to the United States, he resigned his Navy commission to become the first physics professor at the newly founded Case Institute.

SWIMMING ACROSS A STREAM

Michelson's gadget was based on a remarkably simple idea. In ordinary language, it is that *it takes less time to swim across a stream and back than to swim the same distance upstream and back.* This fact is by no means self-evident, so we will establish it here. The required mathematics is rather straightforward and will be useful later in the treatment of relativity.

Suppose a swimmer swims with velocity c and heads across a stream that moves with velocity v. If he wishes to avoid being swept downstream, he must compensate for the current by pointing somewhat upstream. Though he will appear from *shore* to be crossing the stream directly, his motion in the *water,* as seen from a boat drifting with the current, will be along a diagonal, the hypotenuse of the triangle in Fig. 8-1. The other legs of the triangle represent the motion of the stream and his motion as seen from the shore.

This obviously slows him down somewhat. Using the theorem of Pythagoras on the triangle in Fig. 8-1, we see that his velocity as seen from the shore must be $\sqrt{c^2 - v^2}$. Let us now calculate the ratio between his velocities with respect to the water and the land, because it will be very useful later on. It is

$$\gamma = \frac{c}{\sqrt{c^2 - v^2}} = \frac{1}{\sqrt{1 - v^2/c^2}} \tag{8-1}$$

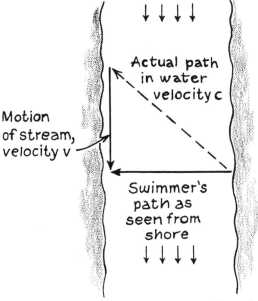

Motion of stream, velocity v

Actual path in water velocity c

Swimmer's path as seen from shore

Figure 8-1

This ratio should be inspected very closely, as it plays a key role in the theory of relativity, where it is designated by the lowercase Greek letter γ (gamma). Because of its importance, the way γ varies with velocity is shown in the graph in Fig. 8-2. As a numerical example, suppose the swimmer can do 5 mi/h in a stream that moves at 3 mi/h. Then $v/c = 0.6$, $v^2/c^2 = 0.36$, and $1 - v^2/c^2 = 0.64$; since the square root of 0.64 is 0.8, we have $\gamma = 1.25$, which means that the swimmer actually swims 25 percent farther than the width of the stream.

Note that if $v = 0$, that is, the stream is stationary, $\gamma = 1$, as we might expect. The formula cannot be used if v is greater than c, for we would then have to take the square root of a negative number. The significance of this in our example is that if the stream moves faster than he can swim, there is no way the swimmer can avoid being swept downstream, so the triangle in Fig. 8-1 is not applicable to the case. Its significance in the theory of relativity is far more profound, as we will see.

The trip up and down the stream is somewhat more complicated, so we shall not derive the result. The swimmer moves upstream with velocity $c - v$ and returns with velocity $c + v$. But he spends more time swimming at the slower velocity, so the average speed for the whole trip is less than c. The slowdown for the round trip is worse than for the cross-stream case. The ratio of average velocities is in fact γ^2, as readers skilled in algebra are invited to verify for themselves.

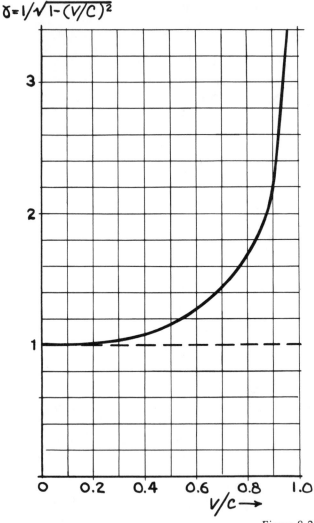

$$\gamma = 1/\sqrt{1-(v/c)^2}$$

Figure 8-2

A RACECOURSE FOR LIGHT

Now we turn to the Michelson-Morley apparatus. The reader has probably anticipated the next step. Replace the stream with the aether wind, the swimmer with a ray of light (which is why we chose the symbol c, the conventional symbol for the velocity of light). But in this case, v/c is not 0.6—it is about 0.0001! How can we measure such a tiny difference?

The answer is that light carries its own yardstick, and a remarkably fine one it is. The "inch" on this yardstick is the length of a single light wave—typically, about 0.00005 cm. What Michelson needed to do was build a device that sets up a race between two light rays, with a way to judge the winner to a fraction of a light wave.

Michelson's racecourse is depicted in Fig. 8-3. A partially silvered mirror transmits half the light falling on it, reflecting the rest, thus setting up two light beams at right angles. Two ordinary mirrors send these back along their paths, returning to the half-silvered mirror, which recombines them. If the waves arrive crest to crest, the center of the pattern in the eyepiece will be bright. If they arrive trough to crest, it will be dark. But how equal are the two paths? Michelson avoided this question by *rotating the apparatus*. Thus, he need never know which light ray actually "won" the race—only that by switching around the upstream and cross-stream paths the margin of victory was changed by some fraction of a light wave.

One might reasonably ask how Michelson really knew which way the aether wind was blowing. While the motion of the earth in its orbit is 30 km/s, perhaps the whole solar system is moving through space in a manner that counteracts this. But if the sun's motion cancels that of the earth in January, it must add to it in July, when the earth is traveling the other way in its orbit.

Indeed, Michelson's measurement was nearly sensitive enough to detect an aether wind from the earth's *rotation* alone, even if all other motions were canceled. And later refinements of the experiment were more sensitive still.

Building a successful apparatus was no mean feat. The racecourse, now called a *Michelson interferometer*, must be capable of being rotated without changing its dimensions by even a small fraction of a wavelength of light. For his first attempt, in Berlin in 1880, he tried using a massive

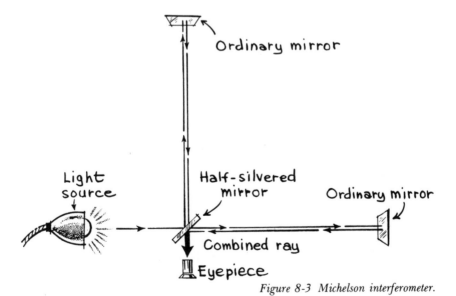

Figure 8-3 Michelson interferometer.

structure of precision-machined steel. Even so, it was so sensitive to vibrations that it could be used only at night, and when an assistant 100 yd outside the laboratory stamped on the ground the effect was visible. In the Cleveland version, the device was mounted on a huge sandstone slab floated on a pool of mercury, with multiple mirrors to lengthen the racecourse to over 30 ft. All was set to go; Michelson had a gadget that could measure the speed of the earth to an accuracy of 1 km/s.

The sandstone slab was set into slow rotation, taking over a minute to turn all the way around. As it went around, each leg of the interferometer took its turn at being the cross-stream path. A quarter turn later, it would be the up-and-down leg. The eyepiece was replaced by a camera looking through a slot that slowed only a small section of a single ring. With the camera shutter open, the film moved automatically as the apparatus turned. Any shift in the ring pattern would have produced a wavy line on the film. Instead, it was straight as a ruler's edge!

The news was very upsetting to a world of physicists convinced they might be the last generation of their kind. Refining, rechecking, repeating, or just trying to explain away Michelson's result became a major activity for several decades.

EARLY ATTEMPTS TO EXPLAIN THE RESULT

The full story of attempts to explain away the Michelson-Morley experiment is interesting and at times amusing, for it still ranks as one of the biggest surprises in the history of physics. Only one such attempt is worth mentioning, however, for it was the one that was on the right track, at least mathematically.

Following a suggestion by the Irish astronomer C. F. FitzGerald, H. A. Lorentz, a remarkably able Dutch theoretical physicist, saw a clue to the mystery of the failure to detect the aether wind in the discovery of the electrical structure of atoms, which came only a few years after the original Michelson result. Seeing that matter was held together by electrical forces, Lorentz speculated, and seeing that it is the aether that transmits these forces, it is just possible that the aether wind "flattens" matter in the direction along the wind as it passes through. It must flatten it by exactly the required amount, γ, to make the Michelson race come out in a dead heat. Of course, no material ruler could detect the effect, for it would conveniently shrink the same amount as whatever it was measuring.

But in the Lorentz theory there remained optical ways of measuring the shrinkage. While a trip across the aether wind would always end in a dead heat with one up and down the wind, the total length of each path in light wavelengths would change as the aether wind changed. Since the speed of the aether wind changes at any point of the earth's surface as the earth rotates, an optical measurement of the length of an object would change with the time of day. Several of the most renowned physicists of the time tried such experiments but failed to observe the contraction.

Meanwhile, Michelson was enjoying to the full the fruits of his success. Had he succeeded in finding the aether wind, he would have gained no mean reputation as a gadgeteer and a careful experimenter. To *fail* to find it, in a convincing manner, and to have a spectacular array of scientific talent out looking for a way to account for the result, made him a real celebrity. Following the accepted academic success formula, he kept moving. First leaving Case Institute for Clark University, he was later lured away to found the physics department of the new University of Chicago, where boy-wonder President William Rainey Harper was using generous libations of Rockefeller money in an attempt to create an "instant Oxford." Sticking to his chosen specialty of large optical instruments, Michelson had a number of celebrated successes. One of the most notable, for sheer physical size, was the mile-long vacuum pipe he set up on the California desert to make a precise determination of the velocity of light.

Despite its cosmopolitanization during World War II, American physics research has retained some of the Michelson flavor. The willingness to build research instruments on a stupendous scale has always been a distinguishing feature of the American style. Many a physicist in this country has carved an illustrious career out of his skill as a gadgeteer without ever personally contributing to physical thought. The giant particle accelerators (popularly known as atom smashers), which are among the most expansive (and expensive) monuments ever erected to human curiosity, would probably have tickled Michelson's fancy.

CHAPTER NINE

Origins of Relativity: How Long is a Moving Train?

But in physics I soon learned to scent out the paths that led to the depths, and to disregard everything else, all the many things that clutter up the mind, and divert it from the essential. The hitch in this was, of course, the fact that one had to cram all this stuff into one's mind for the examination, whether one liked it or not.

—ALBERT EINSTEIN

IOGRAPHICAL SKETCHES of famous physicists contain the phrase "his brilliance became obvious early in his schooling" with monotonous regularity. From this cliché it is necessary to exempt Albert Einstein, one of the handful of twentieth-century physicists whose name has become a household word. In contemporary journalese, Einstein in his late teens would have been described as an undisciplined, confused, middle-class high school dropout.

Einstein was born in 1879 to a family of minor industrialists in Bavaria. When in his mid-teens he withdraw from his Gymnasium (roughly the equivalent of a prep school), his strict schoolmasters breathed a sigh of relief, for the dreamy-eyed woolgatherer was clearly "uneducable." After a year of wandering in Northern Italy, Einstein was informed that his father's business had failed and he must train for some profession. Of his two obvious talents, mathematics and the violin, the former seemed most immediately salable. So he took the entrance examination for the Swiss Federal Polytechnic School in Zurich, then the "MIT" of central Europe and like its American sister usually known by its initials, ETH in German.

Because of his weakness in Latin and Greek, Einstein failed the exam. There was, however, a way to get around this obstacle. Graduates of Swiss secondary schools could be admitted without taking the exam. Furthermore, the school in the town of Aarau had a reputation for dealing kindly with free spirits who bridled at conventional school discipline.

The choice of Aarau was fortunate, as it was run by disciples of the educational reformer Heinrich Pestalozzi, who stressed the development of the visual imagination, young Einstein's strongest point. Einstein later insisted that his best ideas always came to him in the form of visual images, the mathematical and verbal expressions following months or even years later. Aarau proved a comfortable route to the ETH.

His classmates in Zurich have described Einstein as charming and witty but an indifferent student who attended cafés regularly and lectures sporadically. His voracious appetite for reading rarely extended to the books required for his courses. Though his friends were convinced that this charming fellow must possess some sort of brilliance, there were no indications that it was the kind required for success in as disciplined a field as physics, but they did their best to help him get through. Both his friends and the few sympathetic professors rejoiced when he actually managed to graduate in 1900 with an uneven record, standing at the top of his class in a few subjects but with a dubious record in others.

For a year or so he earned an irregular income as a tutor, until the influence of one particularly close friend, Marcel Grossmann, landed him a sinecure as a patent examiner in the Swiss capital of Bern. In this unlikely bureaucratic niche he enjoyed eight of the most productive years in the history of physics. Not only did he develop the theory of relativity, perhaps the greatest single-handed contribution to physics since the time of Newton, but he took some of the most important first steps toward the quantum theory, as well as achieving some of the early insights into the nature of solid matter. He also gave the first exact treatment of the brownian motion, the irregular movement of microscopic objects resulting from the bombardment of moving atoms, an important direct confirmation of the existence of atoms. His first landmark papers on all these subjects appeared in one feverish 11-month period during 1905. It reflects to the credit of the European scientific community that despite his obscurity and questionable credentials, the value of his work was almost immediately recognized. After rushing through the formalities to get his Ph.D., he served from 1909 to 1914 in a succession of professorial chairs in Zurich and Prague, culminating in the creation of a special position, totally free of specified duties, for him at the University of Berlin. He

Einstein during his Bern days. (Photo by Lotte Jacobi.)

looked upon this honor with mixed emotions, for his Patent Office days had been a welcome obscurity that gave him several precious years of freedom from the pressures that normally beset a young academic. The publish-or-perish policy had already become established at Central European universities, and in his later years Einstein was to comment:

> For an academic career puts a young man into a kind of embarrassing position by requiring him to produce scientific publications in impressive quantity—a seduction into superficiality which only strong characters are able to withstand. Most practical occupations, however, are of such a nature that a man of normal ability is able to accomplish what is expected of him. His day-to-day existence does not depend on any special illuminations. If he has deeper scientific interests he may plunge into his favorite problems in addition to doing his required work. He need not be oppressed by the fear that his efforts may lead to no results. I owed it to Marcel Grossman that I was in such a fortunate position.

The most spectacular result of Grossmann's patronage was that while only 25, Einstein promulgated the theory of relativity, which struck even its most enthusiastic early adherents as bizarre.

What Einstein did was to explain the result of the Michelson-Morley experiment by the audacious step of deliberately postulating what appeared to be a paradox and then insisting that this paradox be resolved by a complete reexamination of the concepts of distance and time, notions which had been taken as self-evident and undefinable throughout the history of physics.

THE POSTULATE OF RELATIVITY

The central postulate of relativity is deceptively simple:

> The velocity of light is the same for all observers, in all directions, regardless of their state of rest or motion.

Obviously, the postulate takes care of the Michelson-Morley result by *fiat*. The velocity of the light signal in both arms of the interferometer is assumed to be the same, regardless of the speed or direction of the earth's motion. Rotating the apparatus changes nothing. But the paradox, too, is inescapable. If one observer finds light moving at c, how can another observer moving in the direction of the signal get the same answer? Clearly, if it moves at the velocity of light with respect to one, it cannot so move with respect to the other.

Put another way, the postulate seems to give two conflicting answers to the question: Does light behave like a bullet or like the sound of the gunshot? If a bullet is fired off the front of a moving train, it acquires the velocity of the train in addition to its normal speed. The sound does not, but travels through the air at its usual rate. Thus, a stationary observer would find the sound traveling normally and the bullet traveling faster than usual. If an observer on the train were to perform the measurements, he would find the bullet traveling at its normal speed and the sound moving slower than normal forward and faster than normal backward. What Einstein's postulate suggests is that, to the man on the train, light behaves like a bullet, while to the man on the ground it behaves like the sound of the shot!

Though Einstein's step was a bold one, it was not totally out of harmony with the thinking of others. At nearly the same time, the French physicist-philosopher Henri Poincaré placed a strikingly similar interpretation on the Michelson-Morley experiment. Poincaré noted that such concepts as "absolute" motion and rest had never really had a place in physics, ever since the formulation of the principle of inertia. Perhaps the Michelson-Morley result was just one specific manifestation of a general principle: *no experiment whatsoever can measure absolute motion.* One remains free to regard any body that moves in a straight line at constant speed as standing still. Though unaware of Poincaré's suggestion, Einstein was to carry through its realization, the reformulation of physics so that the propagation of light would not seem to violate this general principle. For it was in the hope that the earth's "absolute" motion could be detected that the experiment was undertaken.

Einstein's insight was to see that while the crucial problem was the propagation of light, the problems it raised were far more general. He examined the paradox in detail to see just what its consequences were, and whether physics could learn to live with them. Before entering into the presentation of his arguments, it is best to begin with a warning: the arguments back of relativity are simple and straightforward, but the consequences, though also reasonably simple, seem to fly in the face of common sense. The normal reaction to a first exposure to relativity is: "I think I understand it; I just don't believe it." Thus, the reader is likely to feel the same frustration as Alice talking to the Red Queen. Normally it takes a physicist about five years of contact with the ideas of relativity to feel comfortable with them—not because they are complex or obscure but just terribly strange.

One rule must be kept firmly in mind, for it will help make the burden lighter. All the peculiar features of relativity arise from a single problem: *when something happens far away, it takes time for the news, traveling at the speed of light, to reach us.* Two observers in relative motion, each applying Einstein's postulate in his or her own way, will disagree about just how long the delay was. Thus they cannot agree on when a remote event took place. Though they agree on what they see, they interpret it differently. We will find that this leads to a number of more bewildering disagreements.

Thus our guideline through the thickets of argument to come will be *to accept disagreements over the time of remote events but to reject any situation that calls for a dispute about actual observations as a violation of the postulate.*

SHIPS THAT PASS IN THE NIGHT

Two of the most direct consequences of Einstein's postulate are that the speed of light becomes the upper limit for all velocities and that observers moving with respect to one another cannot agree on whether two remote events happened at the same time. To show this, we fall back on what Einstein called *Gedankenexperimenten,* or thought experiments. These are not mere pedagogical tricks; in many cases, they are based on the very same visual images that led him to the theory.

We start with a bit of science fiction. Imagine two space ships passing each other in the far reaches of outer space with a relative velocity close to the speed of light. As they meet, let a bright flashbulb go off between their ships. By the postulate of relativity, the crew of each ship has a perfect right to believe that it sits at the center of a sphere of light, expanding in all directions at velocity *c,* while the other ship is winging away close to the edge of that sphere. We are obliged to find a way for both crews to be right.

To start with, it is clear that the relative velocity must be less than *c.* Otherwise, each crew would insist that the other is actually *outside* the expanding sphere and could not possibly have even seen the original flash! Since we do not allow disagreements on actual observations, speeds greater than *c* must be ruled out.

This is the relativistic significance of the rule that *v* must be less than *c* to compute γ in the preceding chapter. Velocities greater than *c* are simply illegal.

Put another way, if there exists *anything* that is so queer as to obey Einstein's postulate, its velocity must be unique and the limit of all possible velocities. There cannot be two separate phenomena with different speeds of propagation both of which obey the postulate.

This represents an enormous change in our attitude toward the velocity of light. No longer is it a property of light itself or of the medium in which light travels (which we have obviously abandoned anyway). It is a fundamental property of the universe, applying to material objects as well as to light. We shall see in the next chapter that the true significance of the speed of light is not that it is an accidental property of electromagnetic waves but that it is a fundamental scale factor in the universe between measurements of distance and time.

It is interesting that though Poincaré was not bold enough to himself create the "new dynamics" for which he foresaw the necessity, he was able to anticipate this feature of it—that "superlight" velocities would be impossible.

At the present point we are in no position to say how this speed limit is enforced. Why can't a material object be accelerated beyond the speed of light? In Chap. 11, when we come to grips with the changes in Newton's laws required by relativity, we shall find the speed limit is self-enforcing.

To return to our space opera, let the two crews agree to turn around and have another go at it and try to resolve the disagreement. Reflecting some of the light back is the only way to tell where the sphere of light is. One crew agrees to rig reflectors far from their ship, at equal distances fore and aft. If the flashes from both reflectors return to the ship at the same time, it must be at the center of the sphere. This is exactly what happens, so the crew of this ship claims to have proved their point. To drive it home they even produce a photograph, taken by a camera with a fast shutter, showing both reflectors at once.

But the other crew has an equally ready explanation of what happened. Since they see both ship and reflectors in rapid motion, the light headed forward must have taken a long time to catch up with the reflector, which was rushing away nearly as fast as the light itself. Its return trip was quicker, with the ship rushing to meet it. The light from the rear reflector did the fast part of the trip first, with a slow return as it caught up with the ship. The round-trip times are the same, regardless of the order of the fast and slow legs.

So this crew concedes that the flashes from both reflectors return to the *ship* at the same time. What they deny is that they hit the *reflectors* at the same time! They concede the validity of the observation but deny that it proves their rival was in the center of the sphere.

Each crew can cling securely to its own interpretation. Since no one can experience directly what is happening in two different places, there is no way to settle the argument. They are both right. The postulate of relativity has withstood its first test, though we have had to discard the idea that the phrase "happened at the same time" has an unambiguous meaning that all observers must accept.

Remember that the expanding sphere of light was nothing more than an abstraction to begin with. No one can experience the whole sphere at once. It can only be reconstructed by observing reflected light and estimating how long ago it was reflected. In our next example, we will find that even a seemingly tangible reality such as the length of a solid object is an abstraction of the same sort.

THE ELASTIC TRAIN

We move now to one of Einstein's own favorite examples, measuring the length of a moving train, which he claimed to have dreamt up while riding to work through the streets of Bern. Of course, in his mind's eye he replaced the slow-moving streetcar with a fast express moving close to the speed of light.

At each end of the train, a conductor is stationed. At the center of the train is a light visible from both ends. When the light flashes, each conductor drops a marker alongside the rails, as shown in Fig. 9-1. Observers on the ground are asked to measure the distance between the markers. But they demur, insisting that the markers were not dropped at the same time, so the measurement is completely invalid!

As they see it, in the lower half of Fig. 9-1, the conductor at the rear of the train was moving toward the flash of light and saw it before his colleague at the front, who was moving away. The marker at the rear was dropped first, and by the time the other was dropped, the train had moved. Thus the markers are farther apart than the length of the train.

One thing is clear; observers on the ground believe the train is shorter than the conductors say it is, since the latter insist that the markers were properly dropped and the postulate of relativity supports both viewpoints equally. There is only one way out of this dilemma: *when a moving object is measured, its length will appear shorter than when it is standing still.*

This was Einstein's analysis of the Lorentz contraction, and we shall

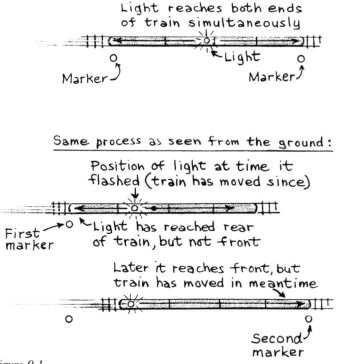

Process as seen by conductors

Light reaches both ends
of train simultaneously

← Light

Marker

Marker

Same process as seen from the ground:

Position of light at time it
flashed (train has moved since)

First marker

Light has reached rear
of train, but not front

Later it reaches front, but
train has moved in meantime

Second marker

Figure 9-1

see in the next chapter that it gives quantitatively the right result for the contraction. But where Lorentz viewed it as a real contraction, Einstein saw it as an apparent one resulting from the difficulty of deciding unambiguously when the two ends of a moving object are where. And the result is not peculiar to this somewhat abstract method of measuring the length of a moving train. A different method will be used in the next chapter. But unlike Lorentz's "real" contraction, there is no way for the contracted observer to see his own contraction. It is not really there for him, but only for the stationary observer.

The moral of this tale is that *a solid object is no less an abstraction than the expanding sphere of light* of the preceding example. No one can experience the whole object all at the same time. If you are standing at one end of a train, your information about the rest of it is old news. This presents no problems unless the train moves or changes in some way. Our faith in the reality of solid objects comes solely from the fact that in our ordinary experience nothing moves or changes much in the time it takes light to get to us.

Einstein exploited this example to make one more point. Suppose the conductors use the light signal to synchronize their watches (in Einstein's time, Swiss railway conductors were reputed to carry the finest watches in the world). The observers on the ground would again object, insisting that the watch at the rear of the train was set first and is thus slightly ahead of the one at the front.

To check this, let the conductors agree to meet at the front of the train and compare watches face to face. The postulate of relativity assures them that the watches were set correctly in the first place, so they must agree when they meet.

This face-to-face check is a direct observation, with no need to correct for a time lag. So the observers on the ground must concede that the watches now agree. Somehow, bringing a watch forward caused it to lose time, eliminating the discrepancy. How can this happen?

Einstein's answer was that *a moving clock appears to run slow,* at least as seen by a stationary observer. The observers on the ground argue that by walking forward the conductor added his motion to that of the train, causing his watch to run slower still. The reason why this is so will be the starting point for the next chapter.

THE FEELING IS MUTUAL

The next example will recapitulate what we've already learned, stressing one important point: the disagreements we have discussed are perfectly symmetric. Each observer claims to be the one at rest, so whatever one says about the other, the other will say the same thing back.

To emphasize this, consider two identical trains, each equipped with synchronized clocks at both ends and in the middle, with an observer stationed by each middle clock. At the instant the centers of the trains pass, depicted in Fig. 9-2, each observer feels as follows:

1. The other train is shorter.
2. The clock at the head of the other train is set slow.
3. The clock at the rear of the other train is set fast.
4. All the clocks on the other train are running slow.

Analyzed in detail, the disagreement is not as irreconcilable as it seems. For example, both agree that when the two clocks at the left end of the picture are compared, *B*'s is reading ahead of *A*'s. They only disagree about which of these two clocks agrees with the ones at the centers of the trains.

The disagreement is completely resolved if one considers what each observer *actually sees* at the instant depicted, which is *not what is shown in the figure*. This is because it takes time for the light signal to reach the center of the train. Each must correct for this time lapse to make a meaningful statement of where the end of each train is and what each clock reads at that instant. And the two observers disagree about how long the time lapse was.

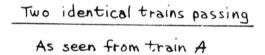

Two identical trains passing

As seen from train A

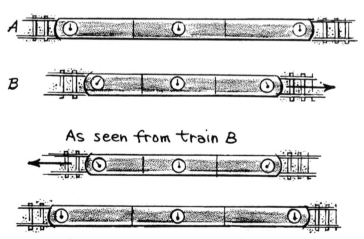

As seen from train B

Figure 9-2

Let us concentrate on the clock at the left end of train *A* and assume that it can be seen by both observers. At the instant shown in the figure, since both are looking at the same clock at the same time, they must agree on what it used to read, not *now* but back when the light they are now seeing left it. How long ago was that? *A* feels the clock is stationary; he feels the time lapse is the time required for light to pass one-half the length of the train. *B,* on the other hand, says this clock is moving away from him. It was closer than one-half the length of the train when the light signal left it, so his correction factor is smaller than *A*'s. When *A* takes into account the larger correction factor, he concludes that the clock agrees with the ones at the centers of the trains. *B* adds a smaller correction, and concludes that the clock is slow compared with those at the centers.

What each saw as he looked toward the left end of the picture was the same: the end of train *B* protruded beyond the end of train *A*. The observer on train *A* feels that the time lapse in the meantime was sufficient to permit the rear of train *B* to pass the end of train *A*. The observer on train *B* feels that the time lapse was shorter, so his train still sticks out beyond the other. Thus, it is just because of ths disagreement over time lapse that it is possible for each observer to feel his train is longer than the other.

THE GARAGE PARADOX

One final and rather amusing example will help to show that while it is true that the contraction is only *apparent,* in the sense that an observer moving with the object will see it at its "true" length, it is nonetheless not quite correct to regard this peculiar effect as merely an *illusion.* Imagine, as in Fig. 9-3, a garage with doors at both ends, set to open automatically just before the bumper hits. Imagine also a car which at rest would be slightly longer than the garage. The car sails through the garage at a speed approaching that of light; it is thus shortened, at least from the point of view of someone in the garage. The rear door opens to admit the car, and then closes behind it before the front door opens to let it out.

But from the point of view of the car's rather reckless driver, it is the garage that gets shortened. As he sees the process, the car must at some point stick out at both ends, and both doors must have been open at once! Common sense tells us one or the other must be wrong—either the car was at one point in a closed garage, or it wasn't!

Again, relativity insists that this is *not* a question with an unambiguous answer. The secret lies in the difference in the sequence of door openings and closings, as seen by the two observers. These are summarized below:

A car passes through a garage at six-tenths the velocity of light. The car is actually slightly longer than the garage

An observer in the garage feels there is an instant when the shrunken car is inside a closed garage...

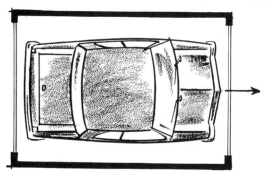

...while the driver of the car, who believes the garage has shrunk, feels quite differently

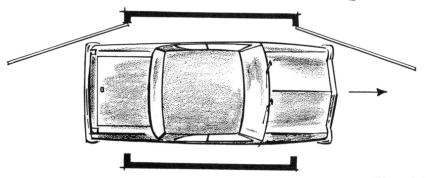

Figure 9-3

Man in garage	Driver
"In" door opened	"In" door opened
"In" door closed	"Out" door opened
"Out" door opened	"In" door closed

The man in the garage is sure he knows in what order he opened the doors, but the driver is equally sure both doors were open at the same time. Again, as in the case of the identical trains in the preceding section, *what they actually see at the instant they are directly opposite one another is the same*. Both actually see the door toward the front of the car closed and the other door open, but both maintain that this represents an earlier state of events, because of the time required for the light to reach the center of the garage. The man in the garage maintains that the back door has closed in the meantime, while the driver maintains that the front door has opened in the meantime. This disagreement is possible because the man in the garage feels that light reaches him from the two ends of the garage in the same amount of time, while the driver, who regards the garage as moving, feels that the light reaching him from the front of the garage left earlier than that from the rear.

Lest this analysis restore the feeling that the whole effect is an illusion, let us emphasize that *for all practical purposes the man in the garage can act as if the shrinkage of the car were real.* That is to say, he can set up the sequence of door openings and closings under the assumption that the car has shrunk, and he will not be contradicted by the splintering sound of a car crashing through a garage door.

The example illustrates how the disagreement as to time sequence leads to a different interpretation of the spatial relations of objects. This is the source of the intimate connection of space and time which is the most striking feature of relativity.

CHAPTER TEN

The Wedding of Space and Time

Alice laughed: "There's no use trying," she said;
"one cannot believe impossible things."
"I daresay you haven't had much practice,"
said the Queen. "When I was your age, I always did
it for half-an-hour a day. Why, sometimes I've
believed in as many as six impossible things before
breakfast."

—LEWIS CARROLL, THROUGH THE LOOKING GLASS

THIS CHAPTER WILL serve to put relativity on a quantitative basis by answering the question: Just how much does a moving clock slow down and a moving object shrink along its line of motion? The arguments themselves will be child's play, but once again the implications of the simple formulas may prove bewildering. The problem is complicated by the fact that three separate but related effects are occurring at once:

1. Moving clocks appear to run slow.

2. Moving objects appear shrunken along their line of motion.

3. The setting of clocks in two different places depends on the motion of the observer.

To cope with these effects, we must find a gedanken experiment that allows us to handle the clock slowdown without worrying about the other two effects. Next we use the result obtained to deal with the shrinkage without worrying about clock settings. The third effect comes from the disagreement about the time of remote events and is easy to calculate. Finally, with all three problems separately solved, we can reunite our view of the universe in the concept of *space-time,* the most "psychedelic" of all of Einstein's ideas. This is the task of this chapter.

HOW SLOW DOES THE CLOCK RUN?

To handle the clock slowdown, we choose for an example a measurement of the speed of light *across* a train rather than along the direction of motion. This is because a stationary observer and one on the train have no disagreement about its width; the position of the sides of the train with regard to its center does not vary as it moves. Both will agree, for example, that a light ray from the center of the train reaches both sides simultaneously. This eliminates the problem of having to deal at the same time with the contraction along the line of motion. To eliminate the third problem, let us use a measurement the *moving* observer can perform with a single clock: let him measure the time it takes for a beam of light to cross the train, bounce off a mirror, and return to its source. Meanwhile, a stationary observer times the same light beam with his own clocks, and each observer reports his results to the other.

It is clear from Fig. 10-1 that the two observers disagree about the distance traveled by the same light ray. Since the man on the train believes it has crossed the train directly, the man on the ground believes it has traveled forward a bit and thus has gone farther. If the train has width w, the man on the train believes it has gone exactly $2w$. How far the man on the ground believes it to have gone was derived in Chap. 8 in the analysis of the Michelson-Morley experiment. It is $2\gamma w$, to be exact, for the path of the light ray as seen from the ground is exactly the sort of diagonal that goes with a cross-stream swim.

Now the Einstein postulate requires that when each observer measures the speed of light, both get the same value. That is, when each divides what he believes to be the length of the light path by the elapsed time read from his clock, the result must be the same number. Since the moving observer uses a smaller figure for the length of the light path, his clock must also have recorded a shorter time than the stationary one. Thus, it must be running *slower,* by exactly the factor of $\gamma = 1/\sqrt{1 - (v/c)^2}$ required to make up the discrepancy.

We can reverse this argument and ask what the observer on the train thinks is happening to the watch of the observer on the ground; i.e., let us regard the observer on the *ground* as moving. Then let us use a ray the observer on the ground regards as going directly across the train. In this case it is our groundling who feels the light has traveled $2w$ while

Path of light ray across train

As seen by observer on train

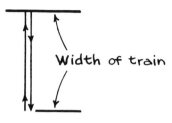

Width of train

As seen by observer on ground

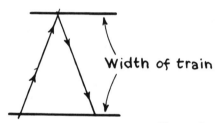

Width of train

Figure 10-1

the observer on the train feels it has gone farther. By the same reasoning as in the preceding case, he must believe it is the watch on the *ground* that runs slow.

If the reader is incredulous (which is proper!) and has the debater's instinct, it is legitimate to ask whether it was fair to swtich light rays between the two situations. This must be done because it is important to use a light ray that *returns to the same place with respect to the observer we regard as moving;* otherwise he would have to use clocks in two different places to measure the elapsed time, and we would have to take into account the difference in clock settings at these two points. The only way to avoid this problem is to insist that whichever observer we regard as moving use only one clock.

If you find it impossible to accept that both observers can get away with believing that the other's clock is slow, keep one important fact in mind: *two observers in relative motion get at most one chance to compare clocks face to face* as they pass. Thereafter they are moving rapidly apart. They can communicate only by light or radio signals that take a noticeable time to get there, and any reply will take longer still. Each estimates the other's clock reading by correcting for this time lag, assuming the other is moving. In Chap. 12 we will see how this permits each to believe the other's clock is slow without contradicting the actual information they receive.

Now that we have the quantitative measure of the apparent slowdown of a moving clock, it is easy to use the result to derive the apparent shrinkage of a moving object along its direction of motion. Again we roll out our familiar train. How are we to measure its length?

THE FITZGERALD SHRINKAGE

The easiest way to measure the length of a train is to measure its speed and then measure how long it takes to pass by a fixed point, such as a telegraph pole. On the pole is an observer with a stopwatch. The observer on the train, looking at his own clock, then says, "Wait a minute—you performed that measurement with a slow clock. Thus, your result must be wrong. My clock says it took longer, and your result is too short."

Now it is the observer *on the train* who must use two clocks to perform the measurement: one at the front and one at the rear of the train. Thus, it is *he* we must regard as the "stationary" observer. Furthermore, since he is the one at rest with respect to the train, it is he who is going to be right in any measurement of its length. The observer up the pole must be getting a value shorter than its real length. The ratio must be the same as the clock slowdown—that famous factor γ.

Remember also that to the observer on the train the telegraph pole is just an abstract point that moves at a known speed from the front of the train to the rear. There is no reason to suppose that this is an invalid way to measure the length of the train. The fact that there is a man on the pole with a stopwatch that looks slow and who therefore finds a shorter length for the train is of no consequence to the observer on the train. It cannot affect *his* measurement in any way.

This is of course only one way to measure the length of a train, but for the others the mathematics are more complicated. Again, the point is that if relativity is correct, this way is as good as any other and will yield the same result.

If the reader is now thoroughly dazed (a common feeling at this stage in the study of relativity) by a world in which moving objects shrink and moving clocks lose time, a numerical reminder may prove reassuring. The most rapid, large man-made objects (interplanetary rockets) move at speeds less than one ten-thousandth of the velocity of light. Thus, γ differs from 1 by about 1 part in 100 million. For more leisurely vehicles, such as supersonic jets, the effect is more like 1 part in 100 billion.

We now move to the disagreement about the time of a remote event. This comes about because each observer feels that the other has moved while the light signal that brings the news of the event was in transit. If one observer believes the event occurred a distance L away, the signal took the time $t = L/c$ to get there. During this time the other observer moved a distance vt. Thus his estimate of the time that the signal was in transit must differ by the time it takes light to travel this additional distance, so we divide this distance by c to obtain the disagreement in time estimates

$$\Delta t = \frac{vt}{c} = \frac{v(L/c)}{c} = \frac{Lv}{c^2} \qquad (10\text{-}1)$$

This same formula gives the difference in clock settings at the ends of a train of rest length L according to an observer who sees the train moving at velocity v.

SPACE-TIME: A FOUR-DIMENSIONAL WORLD

One of Einstein's more profound observations on the significance of relativity was that it removes the separateness of the age-old concepts of time and distance. This view is reflected in the term *space-time* or the more ominous-sounding *fourth dimension* that occasionally filters down into popular speech.

This is illustrated already by our imaginary train. To its passengers, it has a perfectly reasonable length, and time is the same throughout its length. To an observer on the ground, it has shortened but, as if in compensation, has been spread out in time. Time is different at one end of the train than at the other. What to its riders is an object that extends purely in space becomes to others an object with an "extent in time."

The next two steps in this discussion may appear quite arbitrary, but when the conclusion is reached, they will be justified. We wish somehow to compare this time interval with the length of the train. Thus, we must connect it to a length. The most convenient way to do so is to multiply it by c. This is by no means silly—as long as we are going to speak of space and time as unified, we might as well measure them in the same units, i.e., denote a time interval by saying how far a light signal would get in that time. Astronomers have long done the opposite—measured distances with a time-based unit, the light-year. Thus, they state how far a remote star is by stating the time required for light to get here from it.

Our contracted train now has a space extent

$$L_c = \frac{L}{\gamma}$$

and a "spacified time" extent

$$c \, \Delta t = \frac{v}{c} L$$

Now, again at the risk of seeming arbitrary, let us treat these two quantities as if they were two sides of a right triangle. Then we could use the pythagorean theorem to obtain the length of the hypotenuse:

$$\left(\frac{v}{c} L\right)^2 + \left(1 - \frac{v^2}{c^2}\right) L^2 = L^2 \qquad (10\text{-}2)$$

That is, the two combined this way give us back the original length of the train!

The significance of this startling result is illustrated in Fig. 10-2. If we regard time as a *dimension* on a par with the three spatial ones of, for example, north, east, and up, *the combined length of an object in space and time will be the same for all observers.** For an observer at rest with respect to the object, it will be purely spatial in extent; for anyone else, its full extent will be part in space and part in time, but if we measure the disagreement of clocks at the two ends of the object in length units and treat this as a fourth dimension to space, the total space-time length of an object remains the same.

Put another way, what one observer calls a pure *length* another sees as a *combination* of a length and a time interval. Space and time thus lose their separateness, becoming a unified concept.

The point is that the *existence* of an object, those properties of it that are independent of the observer's frame of reference, can only be described in the four-dimensional realm called space-time. To quote a simple but very close analogy, suppose you hold a book up to a bright light and observe its shadow on the wall. Edge on, it produces a thin

---

* To avoid confusing readers familiar with conventional treatments of relativity (others will be totally mystified by this footnote), it must be emphasized that Equation (10-2) is not a relativistic invariant in the usual sense, being composed of a time interval in one frame and a length in another. Its constancy, however, follows from the invariance of the space-time interval involved in measuring the length of a moving object, i.e., the interval between simultaneous position measurements in the stationary frame:

$$L_c^2 = L_0^2 - c^2 \, \Delta t^2$$

The device employed in the argument above avoids the necessity for introducing the negative time metric, which in the intuitive context of the treatment in this chapter may be viewed as a consequence of the relative motion of the two coordinate systems during the time interval in one of them. The author defends this rather unconventional concept as appropriate to the description of a rigid body, rather than the more abstract concept of a space-time interval, which is the more natural and conventional language of relativity.

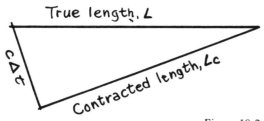

True length, L

$c\Delta t$

Contracted length, Lc

Figure 10-2

shadow. Held with its cover toward the light, it produces a shadow in the form of a broad rectangle. The two-dimensional world of shadows on a wall cannot represent the full solid reality of the book all at once; it exists in its own natural three-dimensional world. Thus, the shrunken "shadows" of moving four-dimensional objects are all we see in our three-dimensional world. The different positions of the paper correspond to the different vantage points of moving observers.

By an interesting coincidence the exactly contemporaneous artistic movement of cubist painting sought a very similar vision of reality. By their own description, the cubists sought to escape the prison of a three-dimensional world of movement and change represented on a static two-dimensional canvas by presenting the multiple views of an active, moving observer. Much of their language describing their art to laymen takes on the same tone as Einstein's similar efforts.

One factor that helps make relativity so baffling is the enormous disparity between our senses of space and of time. The eye can look upon objects a few inches away or gaze upon a grand vista covering tens or at most hundreds of miles. To comprehend greater distances, we must resort to the abstraction of maps. But we can scarcely imagine the time it takes light to cover such distances, which never exceeds a thousandth of a second. Thus our minds are never aware that we are not seeing remote objects *now*.

The word *space-time* should not be taken to imply that relativity makes *time just like space*. Nothing in the theory suggests, for example, a way to move about in time as freely as we do in space. Indeed, it would be more accurate to say that relativity makes space take on some timelike features. Remote parts of space are not accessible to us at the present instant; "right now on Alpha Centauri," our nearest stellar neighbor 4 light-years away, is a meaningless abstraction. *Reality is here and now: the farther from here, the longer from now.* As we look out in space with our telescopes, we are at the same time looking into the past. At their farthest reach, we are gazing into a time billions of years gone, when the universe was young.

But relativity does allow for a peculiar kind of *forward* time travel, the topic of the next section.

THE TWIN PARADOX

One of the most startling predictions of relativity is illustrated by the following science fiction story.

A young astronaut takes a trip to a star 25 light-years away in a spaceship that can travel at 99.98 percent the speed of light. The astronaut has a twin brother, who remains home on earth. Fifty years pass, and the earth-bound twin, a grandfather bent with age, goes to the spaceport to welcome his adventurous brother. The astronaut bounds down the gangplank, because for him only one year has elapsed, and he is still young and vigorous!

From the point of view of the twin on earth, this is because time slowed down on the ship. Clocks, and biological aging processes, are slowed to one-fiftieth their normal speed on this fast spaceship. To the astronaut, the round trip only took a year, because the distance shrank to $\frac{1}{2}$ light year. But both agree that the astronaut is now 49 years younger than his twin!

Early in the history of relativity, this story was offered as a refutation of the theory. Why isn't the twin *on earth* the younger? After all, from the point of view of the astronaut, it is the *earthbound* clock that ran slower! There appears to be a contradiction.

The answer is that one *can* make a distinction between the astronaut and his brother. The astronaut had to leave the earth, accelerate to a stupendous speed, and turn around (another period of acceleration) at the star. Thus, he is not in uniform motion at constant speed, and the symmetry of relativity, by which he feels the earthbound clock runs slow, does not apply. Analyzed in detail, the problem reveals that from the point of view of the astronaut, most of the 50 years passed on earth during the short time he was turning around at the star.

That this is the resolution of the paradox can easily be seen if we imagine that, at the star, there is a clock set to "earth time." Since the astronaut regards this clock and one on the earth as moving clocks, they are indeed running very slow. But since the one near the star is far back along the line of apparent motion, it is also set ahead of the earth clock. In fact, it reads nearly 25 years ahead.

When the astronaut reaches the star and fires his rocket to stop and then turn around, he reverses the situation. The two "earth time" clocks are now headed in the opposite direction, with the one on earth trailing. Thus, it is now 25 years *ahead of,* rather than behind, the clock at the star. But the astronaut is close to that other earth-time clock. In the short time he spends turning around, very little time elapses on it, and it certainly doesn't run backward. If the remote clock on earth switched from being almost 25 years behind to almost 25 years ahead, nearly 50 years must have passed on earth in the brief time he spent turning around. Thus, the astronaut agrees with his twin brother: more time has passed on earth than on the spaceship, and he is now some 49 years the younger of the two!

This 50-year "leap" in earth time is of course an abstraction, a matter of the astronaut changing the way he interprets remote events because of his change of direction. While it happens, he is far from earth and has no direct knowledge of what is going on there. In Chap. 12, we will look into this process in more detail, for it represents one of the loose ends that drove Einstein to go beyond the simple version of relativity we have studied so far.

The twin paradox was a controversial subject in relativity until a few years ago, with a few physicists who otherwise accepted the theory denying the validity of the argument presented here. But experiment has settled the question. In 1972, a scientist from the U.S. Naval Observatory, which is responsible for maintaining our time standard, took a round-the-world trip on a scheduled airline. As traveling companions he took along four cesium-beam *atomic clocks,* which over the time span of such a voyage can be trusted to a few billionths of a second. These clocks indeed lost time when compared with identical ones that remained in Washington, and the time lost agreed exactly with Einstein's prediction. It was a rather cheap experiment, by present-day standards. The total cost was two airline tickets (one for the clocks!).

If nothing else, the twin paradox should be the final convincer that relativity concerns more than illusion, for there is nothing more concrete than returning from a trip to find yourself younger than your twin—or even than your own children!

The literary possibilities of this phenomenon have been fully exploited by science fiction writers. One series of such stories visualized an age when humanity has spread to habitable planets scattered throughout the galaxy, a civilization spanning thousands of light-years. In their powerful, speedy spaceships, a breed of astronauts maintains the skimpy "commerce" of this vast civilization, condemned to a strange existence in which they return to a familiar port only after centuries or millennia have elapsed, thus enjoying a peculiar sort of alienated immortality within the normal three score and ten.

CHAPTER ELEVEN

$E = mc^2$ and All That

W E ARE ABOUT TO reach a turning point in the discussion of relativity. So far, the problem has been: What do we have to pay for believing in Einstein's postulate? The arguments have carried the force of necessity: if we want to keep the postulate, we must learn to put up with a lot of outrage to our common sense. We have no way out of radically revamping our notions of space and time.

Since these concepts are the whole basis for the description of motion, the reader should by now have the uncomfortable suspicion that the whole science of mechanics may now be on mighty shaky ground. Must we discard three centuries of thought, sweep away Newton's laws, energy and momentum conservation, and the like? It turns out that the answer is no. Surprisingly, most of the edifice of classical mechanics emerges from this part of Einstein's revolution intact.

Intact, yes, but by no means unchanged. And the major changes center on the concept of mass. Newton's first and third laws are obviously untouchable; unless we insist that mechanics concern itself with the problem of change of state of motion, and that such changes come about through mutual interactions of bodies, nothing recognizably newtonian would survive. We will find that momentum conservation and the second law can be rescued merely by allowing mass to depend on velocity in a way that is unimportant except at speeds approaching that of light. Moving on to energy conservation, we will find that this result leads to the duality between mass and energy expressed in the most widely publicized formula of twentieth-century physics, $E = mc^2$.

But the arguments in this chapter should be far less of a strain on the imagination than those in the preceding. Though they will still appear arbitrary, their consequences will not seem so outrageous.

THE MASS INCREASE

The first task is to establish the dependence of mass on velocity. The example of choice is a grazing collision between two identical bodies, familiar to all billiard players. The moving body is hardly deflected and loses little speed, while the struck body comes out at a slow speed nearly at a right angle to the original motion. How this looks to two observers, each of which was at rest with respect to one of the bodies before collision, is shown in Fig. 11-1. We will concern ourselves only with that portion of the motion of each body which is at right angles to the original line of motion.

There are two reasons for choosing this example. The first is that it is possible by this device to have two observers in relative motion at a high velocity, yet the part of the motion of the bodies we are studying is at a low velocity. The second is that this motion is perpendicular to the high-velocity motion. As pointed out early in the preceding chapter, in this direction there are no disagreements in measurements of length.

We now ask what happens if we assume that momentum is conserved in the perpendicular direction. This means that the product of the mass and the perpendicular component of the velocity must have the same value for both bodies. It does not matter that for the initially

Collision as seen by observer at
rest with respect to body a before
the collision   (observer A)

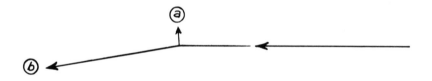

Same collision from point of view of observer
at rest with respect to b (observer B)

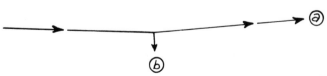

Figure 11-1

stationary body this is essentially the whole momentum while the other body has a much larger component along the original line of motion. Ever since Galileo we have had the right to study components of motion in perpendicular directions separately.

Not only must both bodies have the same perpendicular momentum for each of the observers, but the actual *value* of these momenta must be the same for both observers. Since the bodies are identical, the situation shown in Fig. 11-1 is completely symmetric. Thus, for example the perpendicular momentum of body *a* is the same whether observer *A* or observer *B* actually performs the measurement.

Now to measure momentum we must measure velocity. Let both observer *A* and observer *B* measure body *a's* perpendicular velocity. From *B's* point of view, *A* is using a slow clock. Since they agree on the distance part of the measurement, using a slow clock must give a higher value for the velocity: *A* feels the body goes the same distance in less time.

But then if the *momentum* is to be the same, *A* must be using a smaller *mass*. And since it is *A* who is nearly at rest with respect to body *a,* it is he who must be more nearly right.

The conclusion is that *momentum conservation can only hold if a fast-moving body appears to have more mass than a slow-moving one.* Since it is the same old clock-slowing factor that is involved, the mathematical statement of this rule takes the form

$$m = \gamma m_0 \tag{11-1}$$

where m_0 is the *rest mass,* the mass of a body as measured by an observer at rest with respect to that body.

NEWTON'S LAWS AND THE RELATIVISTIC SPEED LIMIT

Now what of Newton's second law? In Chap. 3 we showed that once one agrees on a method of defining mass, the second law is merely the quantitative definition of force. Though the definition of mass is now somewhat complicated by the necessity for specifying the velocity at which it is measured, it still gives us a unique value for mass. But since now the mass of a body changes as it accelerates, we can no longer write the second law as

$$F = ma \tag{11-2}$$

because both mass and velocity change. We must return to Newton's original statement: force is the rate of change of momentum. At low speeds, $F = ma$ is nearly correct, for the mass changes very little, so all the increase in momentum comes about through changing the velocity. At speeds approaching that of light, the body becomes more and more massive. Then the increase in momentum involves a large change in mass and a small change in velocity.

This property of mass also assures us that Einstein's requirement that nothing move faster than light will never be violated by a material object. As a body approaches the speed of light, the application of a force in the direction of motion will only make it more massive and not change its speed appreciably.

For example, consider a body moving at one one-hundredth of the velocity of light, already a fantastic speed by terrestrial standards. If we let a force act on it until its momentum is doubled, its mass only changes by 0.015 percent, while its velocity very nearly doubles. If, on the other hand, we double the momentum of a body already moving at $0.99c$, we then will nearly double its mass while its velocity will increase only to $0.997c$.

ENERGY AND MASS

The next logical step is to ask what happens to the conservation of energy. Clearly, energy needs to be redefined. For example, a falling body that is given an initial downward thrust at nearly the velocity of light will have its mass increased, but there will be very little change in velocity under the influence of gravity. The logical starting point for the discussion is to consider the *work* done by a force. In the preceding example, the change in momentum was the same whether the result was a large change in velocity and a small change in mass, as for a slow object, or a large change in mass accompanied by a small change in velocity, as for a fast one. But in the latter case much more distance is covered in the process, because the body is moving much faster. Since work is force times distance, *much more work is required when the change in mass is large.*

Though the computation is difficult, it can be shown that the change in mass is just exactly proportional to the work done; thus, the energy of a body is merely proportional to its mass, and kinetic energy would not depend on velocity were it not for the fact that mass depends on velocity. But energy conservation, as has been emphasized, is not merely a mechanical law. Therefore, let us establish the relation between mass and energy by the simpler and more universal device of insisting that light itself obey momentum and energy conservation.

As we saw in Chap. 6, electromagnetic forces can only conserve momentum if we ascribe momentum to the field (and hence to light) itself. If an object emits light in one direction, in order to conserve momentum it must itself "recoil" in the opposite direction. Maxwell established a formula giving the momentum carried by light in terms of its energy:

$$p = \frac{E}{c} \qquad (11\text{-}3)$$

If we stick to the definition of momentum as $p = mv$, we can associate a "mass" with a flash of light:

$$m = \frac{p}{v} = \frac{p}{c} = \frac{E}{c^2}$$

which leads to the famous formula $E = mc^2$.

Is this mass merely a mathematical fiction? Consider the situation depicted in Fig. 11-2. Inside a closed box that is free to move, a flash of light is emitted at the left end. If we insist that this process conserve

Closed box at rest...

begins to move to left when light flashed at left end

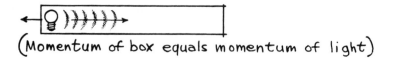

(Momentum of box equals momentum of light)

Stops when light absorbed at right end...

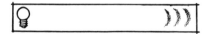

Thus light has transferred mass from one end of the box to the other

Figure 11-2

momentum, the box must recoil ever so slightly to the left. Then let the light flash be absorbed at the right end. The momentum of the light is also transmitted to the box, and it is brought to a halt. Without any external force acting, an initially stationary box has been moved to the left. Since momentum conservation insists that the center of mass of the box can't move as the result of internal forces, as was shown in Chap. 2, the result is exactly as if some real object had moved to the right inside the box.

One might reply that this merely proves that light can be used to transfer mass, not that there is any general relation between mass and energy. But consider the state of the box after it stops. The light no longer exists. None of the actual material of the box has been removed from the left and transferred to the right, yet mass has been transferred between the two ends! What other change in the state of the box must have taken place? What evidence remains of the transfer of the light?

The answer is that absorbing the light has heated up the right end of the box. Conversely, if energy was conserved when the light was produced at the left end, the law of conservation of energy tells us some energy has been removed. For example, the source of energy for the flash might have been the chemical energy in a battery.

What is the net result of the process? Some chemical energy at the left end of the box has been transformed into heat at the right end, and mass has been transferred from the left to the right! Thus chemical energy and heat must also obey a mass-energy relationship. A discharged battery must be lighter than a charged one, a hot object heavier than the same object cold. But since heat is nothing but the energy of motion of molecules, this too must obey a mass-energy relationship. And through the law of the conservation of energy, the mass-energy equivalence can be extended to any form of energy whatsoever.

It must be emphasized that $E = mc^2$ is the *one and only* formula for energy in relativity. What then has become of our old definition of kinetic energy, $\frac{1}{2}mv^2$, which still ought to work at low speeds? The answer is that the kinetic energy merely represents the tiny increase in mass of a slow-moving body. For readers familiar with the calculus, this can easily be verified from Eq. (11-1).* At very small velocities, the additional mass is $v^2/2c$ of the rest mass. Thus, by the formula $E = mc^2$, we obtain $\frac{1}{2}m_0 v^2$ for the added energy. This formula only works if the speed is quite small compared with that of light; it is no longer the *definition* of kinetic energy but merely an approximate formula valid at very small velocities.

---

* Do a Taylor's expansion of the formula at $v = 0$. The term $v^2/2c^2$ is the second term. See the section of the Appendix on this chapter.

THE MEANING OF $E = mc^2$

Since Hiroshima, in the popular media, the formula $E = mc^2$ has become associated with nuclear energy. At the risk of repetition, it must be stated emphatically that it applies equally well to all forms of energy; it is a quite universal and unique formula, as valid for a bonfire as for a nuclear weapon. The only distinction of nuclear energy is that it is the one energy source powerful enough for the changes in mass to be really substantial. Heating water from its freezing point to its boiling point only increases its mass by about one part in 10^{11}. In an ordinary chemical reaction, such as a fire, the combustion products are lighter than the fuel and oxygen used in the fire by about one part in 10^9. Such small changes are just beyond the range of measurement. But in the more violent nuclear reactions, mass changes of about 1 part in 1000 can take place. This is easily within the range of accuracy of mass measurements. From a table of nuclear masses, a physicist can use Einstein's formula to predict the energy release of a previously unstudied reaction.

The crucial factor for the development of nuclear energy and nuclear weapons was the discovery of the nuclear fission chain reaction, one where each disintegrating nucleus triggers several neighbors. The formula $E = mc^2$ by no means predicts the existence of such a reaction, nor is it essential to the understanding of the fission process. Einstein himself played no role in the discovery of fission or in the development of the atomic bomb, aside from the initial one of providing a letter of introduction for the scientists who approached President Franklin D. Roosevelt with the proposal for the project. Had relativity not yet been discovered, it would probably have hampered the efforts of the Manhattan Project very little.

The formula is sometimes mistakenly referred to as a formula for the conversion of energy into mass. It is more than that; *it is a statement that, for all practical purposes, the two are identical.* If you want to know how much energy is in a system, measure its mass.

For example, if a bonfire occurs in a sealed box, insulated so that the heat cannot escape, no change in its weight can occur. Despite the transformation of chemical energy into heat, which represents kinetic energy of the molecules, no change in mass has taken place. If we allow the heat to escape, however, the box will become slightly lighter.

As a last example to illustrate the generality of the mass-energy equivalence, it should be pointed out that it applies equally well to potential energy. A compressed spring weighs more than the same spring relaxed. Whenever a force binds two objects together so that it would take the addition of energy to separate them, their combined mass is less than the sum of their separate masses. The negative potential energy appears as a defect in mass. This is the real source of nuclear energy: the potential energy of the powerful forces that bind the component particles of the nucleus together. A rearrangement of a large, loosely bound nucleus into smaller, more tightly bound ones increases the strength of the binding, lowering the mass of the nuclei. Since this mass is associated with the *field* of force, it can be thought of as distributed in space; indeed, the very presence of a field in a region of space implies energy "stored" there, and therefore mass. Thus, as the concept of field matures in physics, it tends to seem more and more real.

The leftover mass is the equivalent of the change in potential energy. A typical nucleus weighs about 1 percent less than the combined mass of its component particles. The atomic bomb relies on the fact that the heaviest nuclei weigh about 0.1 percent more per constituent particle than nuclei near the middle of the table of elements and on the existence of a process (fission) that makes it possible to trigger certain heavy nuclei into splitting into two smaller parts.

The energy-mass equivalence once again illustrates the role of the velocity of light as a conversion factor between quantities man originally regarded as distinct. Had the equivalence of mass and energy been understood from the outset, there might not have been separate units for the two. Today, physicists working with subatomic particles, where grams are horrendously large and awkward units, use energy units for mass as a matter of course. Viewed in this context, the formula $E = mc^2$ has no more profound significance than the conversion formula 1 km = 0.62 mi. This interpretation of the role of the velocity of light in the mass-energy relationship is similar to its role in the space-time relationship in Chap. 10. Both are motivated by the same spirit, a pervading one in modern physics: the desire to eliminate arbitrariness as far as possible. The idea is essentially metaphysical; it expresses a faith that there are very few or perhaps ultimately no arbitrary fundamental constants in nature, such as the velocity of light. Many attempts have been made in recent years, most a bit premature, to eliminate all such constants or to interpret their meaning on purely mathematical grounds.

The dependence of mass on velocity was one of the first predictions of relativity to be experimentally confirmed. Electrons are so light that it is rather easy to accelerate them to considerable velocities. Those in a typical TV picture tube travel at nearly one-fourth the velocity of light and have a mass nearly 3 percent greater than when standing still. And even higher velocities are easy to obtain. Measurements of the mass of fast-moving electrons were performed as early as 1906 and bore out Eq. (11-1) exactly. The energy-mass relation in nuclear reactions was confirmed to high precision in 1932 in the first artificial disintegration of a nucleus.

The other well-verified aspect of relativity is that of the time dilation. The American physicist Herbert Ives confirmed it in 1938 by measuring the frequency of light emitted by fast-moving atoms. Interestingly, Ives was himself one of the last prominent opponents of relativity.

Today, experimental confirmations of all aspects of relativity are commonplace. Physicists studying subatomic particles work daily with objects traveling close to the speed of light. In particle collisions, these particles bear out all details of Einstein's predictions. For example, some of them are highly unstable and break up spontaneously in the time it takes for a light signal to travel several centimeters. Yet, at close to the velocity of light, they can be transported many meters with no difficulty, because of the slowdown in their internal "clocks." From the point of view of the particle, it is not the slowing of time but the contraction of the laboratory that is responsible for the effect. Yet the end result is the same: it reaches the detector. And when a particle that would be very light when standing still is brought up to a high speed, it acts like a heavy particle in collisions. Finally, the extra mass obtained by accelerating a particle close to the speed of light can be used to produce new particles not previously present.

The final triumph of relativity was to complete the work of Faraday and Maxwell in uniting electricity and magnetism. Einstein was able to show that a magnetic field appears when a purely electric field is seen by a moving observer, and an electric field appears when a purely magnetic one is seen from a moving vantage point.

AN INCOMPLETE REVOLUTION

Though revolutionary in conception, the end result of Einstein's relativity was to save classical physics by reformulating it in a manner consistent with the known properties of light. But this only holds for that part of the theory we have discussed so far, which Einstein called *special relativity*. Like many a revolutionary dissatisfied after his triumph by the incompleteness of his revolution, Einstein moved on in 1916 to a bolder step, called *general relativity*. In this theory he attempted to analyze the problem of light as observed not by uniformly moving observers but by accelerated ones. At the same time he attempted to bring gravitation within the scope of his theory.

The end result was a complete reformulation of mechanics in which the concept of force, for all intents and purposes, disappears altogether. Instead, the action of fields is to distort the very fabric of space itself, until a straight line is no longer the shortest path between two points.

It is this theory that gave birth to the oft-cited (but decidedly false!) legend that only six people in the entire world could fathom what Einstein was talking about. This legend grew because to extract numerical predictions from the theory one must employ some formidable mathematical tools familiar only to specialists in this area. There were dozens of them in Einstein's time, and there are hundreds today.

But the basic ideas of general relativity can be grasped without resort to the arcane mysteries of noneuclidean differential geometry. That will be the task of our final chapter on relativity.

CHAPTER TWELVE

Did God Have Any Choice?

*The most incomprehensible thing about the
universe is that it is comprehensible.*

—ALBERT EINSTEIN

T HE GENERAL THEORY of relativity is first and
foremost a theory of *gravity*. But a closer examination of the twin
paradox provides a handy bridge from the special to the general theory.

To add another dimension to the paradox, imagine that the
astronaut tries to ease the burden of his loneliness by keeping track of
events on earth via radio and TV. This will reveal how complete his
isolation is and how it is possible for him to conclude that it is on earth,
rather than on his ship, that time has slowed to a crawl.

THE LONELINESS OF DEEP SPACE

To understand the astronaut's plight, let us first remind ourselves of the
numerical details of the voyage. His speed is 0.9998c; that is, his velocity
differs from that of light by only 1 part in 5000. At this speed, γ is 50.

These numbers dictate that on the outward half of the trip, *the
astronaut can only receive programs broadcast during the first 1.8 days following
his departure!* It is instructive to examine the contrasting ways in which he
and his brother explain this fact. From the point of view of the twin on
earth, it is a simple matter of a very close race. Since a radio signal can
only gain on his brother by 1 part in 5000, only those broadcast in the
first five-thousandth of the trip can beat him to the star. The arithmetic
is simple: 25 years × 365 days ÷ 5000 = 1.8 days. The programs are so

stretched out in time that the brother must record them and play them back speeded up to make any sense of them.

But the postulate of relativity entitles the astronaut to believe that *earth is rushing away from him, and the star toward him,* at nearly the speed of light. Thus, as shown in Fig. 12-1, the program that meets him at the star must have been broadcast *halfway through the voyage,* or 3 months into it, by his clock. This proves that it is earth's time that has slowed by a factor γ, for 90 days $\div$ 50 = 1.8 days! Thus we see how each brother can justify the claim that the other has the slow clock.

On the return trip, however, the situation is reversed. From his brother's point of view, the astronaut will run back through the rest of 50 years of programs. From his own standpoint, however, it is now the earth and the radio signals that are in a close race. To get ahead by 6 light-months, for that is how far away he believes earth to be, the signals must have been in transit for 5000 $\times$ $\frac{1}{2}$ year = 2500 years! He then remembers that earth time is slowed by a factor of 50 and realizes that this represents 50 years on earth. Thus this 1.8-day-old program must be the forerunner of 50 years' worth to come, *already winging their way toward him.*

Here we have the celebrated 50-year jump. Approaching the star, the astronaut believes earth time to be 25 years *behind.* Moments later, in the same place but traveling in the opposite direction, he has changed his point of view and expects to be greeted by a brother 50 years older than when he last saw him! The *information* is the same; it is his *interpretation* of it that has so radically changed.

Is the astronaut's position self-consistent? Obviously not! He knows for a fact that he was himself on earth only 6 months previously, by his own clock. But as he watches a program sent after his departure, which might even include a replay of his blastoff, he concludes that by this same clock the program is 2500 years old!

The problem is that the astronaut, by reversing direction, has changed what we have heretofore loosely referred to as his *point of view.* To resolve the problem, we must first refine this term into its more exact scientific equivalent, *reference frame.*

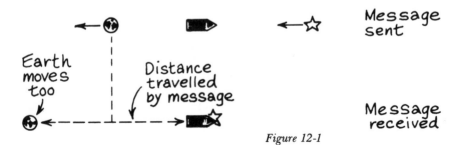

Earth moves too

Distance travelled by message

Message sent

Message received

Figure 12-1

FRAMES OF REFERENCE

The term *reference frame* denotes a system for measuring position and time. Einstein, in his first paper on relativity, objectified it by imagining a grid of rigid measuring rods, with a clock at each junction of the grid.

What relativity says about reference frames is that when you sit in one, rods aligned along the direction of motion in any other frame moving with respect to yours will appear shorter than the others and the clocks will appear out of time with each other and running slow. Thus two observers using different reference frames construct different descriptions of the same process.

What the astronaut did was to *change reference frames in midvoyage.* This forced him to try to merge two conflicting descriptions into one consistent whole. The so-called 50-year leap is *not a real phenomenon but an artifact of this doomed effort.* His brother is in no such predicament. It is possible to make a distinction between the two because in both newtonian physics and special relativity, *not all reference frames are equally valid.* One must distinguish between frames that are moving at *constant velocity* and frames that are *accelerated,* such as the rocket with engines firing, in Fig. 12-2.

It is easy to check that the rocket is in fact an accelerated reference frame, for any loose object in it will "fall" toward the tail, with uniform acceleration. Of course, in newtonian terms, no force is acting on the object; it is really the rocket that is accelerating. Newton's laws do not work in this reference frame, because objects seem to accelerate when no force is acting on them.

So we could let the astronaut's twin say, "Aha! You can't deny that acceleration. You switched reference frames, so how do you expect to be self-consistent? You must accept my version of the story." This is in fact how some books on relativity resolve this question, because to this day many physicists take special relativity as gospel but dismiss the general theory as mere conjecture.

By now, you should have enough of a feel for Einstein's style of thinking to realize that he would never let matters stand at that. Should the astronaut be forced to forswear his right to a self-consistent picture of nature just because he was unfortunate enough to spend some time in an accelerated frame? Why not a physics that works in any reference frame? Besides, are we really all that sure that it is the rocket that is accelerating?

This led Einstein back to a puzzle that had plagued Newton and his successors. The principle of inertia makes velocity *relative,* but the second law makes acceleration *absolute.* Put another way, there is no physical test that can tell whether an object is moving or standing still, but the second law seems to imply that you can tell whether its velocity is changing; there must be a force present. This compelled Newton reluctantly to fall

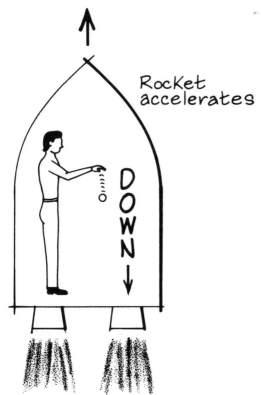

Figure 12-2

back on the notion of *absolute space,* in which you *cannot tell whether you are moving but you can tell whether your motion is accelerated.* Philosophers from Berkeley to Kant denounced this as a monstrosity that must at all costs be expunged from newtonian physics.

What no one had noticed in the intervening two centuries was that Newton *had* indeed dealt with a case where objects accelerate with no evidence that a force was present *except for the acceleration itself.* For what else is gravity?

Inside the rocket in Fig. 12-2, all objects fall with the same acceleration, just as in a gravitational field. *Just as in a gravitational field!* To Einstein, this had to be more than mere coincidence.

This is the insight that led straight to the general theory. Let the astronaut reply, "What accelerated frame? I was in a gravitational field!" His brother will reply, "But you fired your rockets." He can come back with, "Yes—did you want me to *fall* in that gravitational field?" There is nothing in Fig. 12-2 that can tell the astronaut for sure that the ship is really accelerating.

THE PRINCIPLE OF EQUIVALENCE

Like the special theory, general relativity is based on a single postulate that is easy to state in plain language:

> No experiment whatsoever can distinguish a gravitational field from an accelerated reference frame.

As with the special theory, this seemingly innocent statement, which goes by the name *principle of equivalence,* turns out to have mind-boggling consequences. Before we explore them, let us spell out our destination and the route we shall take to it.

Our conclusion will be that *there is absolutely no need for a force of gravity.* The acceleration of falling bodies or planets is simply a case of inertial motion, of objects coasting along on the shortest paths available to them. But these paths are not straight lines, *because space-time itself is curved.*

The problem with these words is that we can scarcely imagine what they mean. It is hard enough for us to accept the fact that our earth is curved. Looking at a flat map, it is hard to visualize that the curved paths we call great circle routes or, to use their mathematical name, *geodesics,* are really the shortest routes between points on the earth's surface.

But at least we can be reassured that the earth sits in a perfectly flat space, and if we could tunnel through it, *that* would be the shortest route. But space itself curved? What do the words mean?

We shall try to remove some of their mystery by using our tried-and-true device of gedanken experiments. These are designed to show in turn:

1. In an accelerated reference frame, the statement is quite trivially true.

2. Light moves in curved paths in accelerated frames and thus also in gravitational fields.

3. If we accept the above as meaning that space-time itself is curved, we can explain everything we know about gravity without ever having to mention a force.

4. Gravity affects time, turning the astronaut's 50-year leap into a real effect.

5. People who live in round worlds but insist they are flat are bound to invent forces like gravity.

6. There are ample experimental tests of all of the above, and in particular one important effect that can be explained in no other way.

The payoff for climbing this path will be a physics that is starkly simple, contains some fascinating curiosities like black holes, and gives insight into the origin of the universe itself.

The rules of the road are simpler than for special relativity. We must simply accept that once we have demonstrated that something happens in an accelerated reference frame, it must also happen in a gravitational field.

THE WARP AND THE WOOF

Galileo could have drawn Fig. 12-3 if he had known about graphs, which were invented by Descartes. But it is more than a graph; in our new four-dimensional language, it is a map of a two-dimensional slice of space-time, showing the path of an object moving freely in the accelerated reference frame of Fig. 12-2. It is obviously a curved line.

Newton would say, "You ninny! What do you expect when you use illegitimate reference frames?" Einstein would reply, "There are *no* illegitimate reference frames. In this one, the path of an object obeying the law of inertia (i.e., with no forces on it) is a curved line. It would be exactly the same in a gravitational field."

So far, we have seen nothing startling. But suppose we are inside the rocket in Fig. 12-2 and shine a light across it from wall to wall. Since the opposite wall will speed up in transit, the light will hit that wall lower than the point it was aimed at, and the path of light across the rocket will in fact be a parabola. The principle of equivalence forces us to concede that the same thing must happen in a gravitational field.

The curvature is so small that we would have a hard time proving this. Suppose that on a very clear day, we shine a laser beam from one mountain peak to another, 20 mi distant. It would get there in a ten-thousandth of a second. In that short a time, a falling body only drops about 10^{-7} cm, a distance not much larger than the diameter of an atom! We could never distinguish the path of a beam that curved *that* little from a straight line.

There is nothing here that would have caused Newton to lose a night's sleep. He would have happily conceded that gravity could affect light. The real argument is about how we describe the *cause* of what we see. So let us move on to an ordinary falling body and see how Einstein accounts for its motion.

You may now object that such a slight curvature could hardly account for the motion of falling bodies. But we must remember that the curvature is not of *space alone* but of *space-time*. The fact that we do not notice the curvature is another consequence of the disproportion between our sense of space and our sense of time.

If you drop a coin from waist height, it reaches the floor in about a half second. While we regard this as a small interval of time, its space

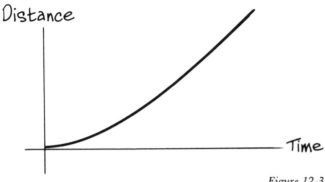

Figure 12-3

equivalent is immense, for in that time a light beam covers about 100,000 mi. The equivalent of Fig. 12-3 for that coin is a graph with vertical dimensions of a few feet but horizontal dimensions of 100,000 mi. We have sampled a huge two-dimensional slice of our four-dimensional space-time, and its tiny curvature is quite sufficient to explain what we see.

Einstein says that when the coin is released, it simply follows its natural curved path in space-time. This brings it, after its long journey, in contact with the space-time path of a point on the floor. This other space-time path is straight, because gravity is not the only game in town. Whatever gives solid matter its hardness must also be able to influence the geometry of space-time. It can overcome gravity and straighten out the floor's space-time path.

Next consider the orbit of the moon. To convert it into a space-time path, we must insert the time dimension. But the space equivalent of a month is so huge that if we computed it, the numbers would be meaningless. So we will resort instead to a scale model.

Reduce the diameter of the moon's orbit to that of an ordinary drinking straw. The length of the straw represents the time dimension. The moon's space-time track is a long lazy spiral circling the straw. We sketch it by first drawing a straight line along the length of the straw and then giving the straw a single twist, so that the line becomes a spiral. The straw would have to be *10,000 mi long!* The curvature of the path is imperceptible but enough to bring the moon once around the earth in a month.

There is no quarrel between Newton and Einstein over the description of these space-time tracks. What they disagree about is their significance. Newton says the tracks are curved by the action of a force. Einstein insists that no force is necessary: space-time itself is curved.

We have now completed the first three (and the easiest) steps of our argument, and still we have no way to choose between the newtonian and relativistic explanations of gravity. The last three will rectify this situation.

DOES ANYBODY HAVE THE RIGHT TIME?

We are now ready to resolve the quandary of our astronaut. Figure 12-4 shows an accelerating rocket ship, equipped with a clock in its nose that sends time signals 10 times each second, to be received further down in the ship.

But the rocket is accelerating, so by the time the signals reach the receiver *it has picked up speed.* Let us say it picks up 0.1c in the time it takes the signals to reach it. Then it will move toward them fast enough to pick up one additional signal each second. We then have *a clock sending 10 signals each second, but a receiver picking up 11!*

How is this possible? Surely we are not *manufacturing* additional signals along the way! We have only one way out, and that is to admit

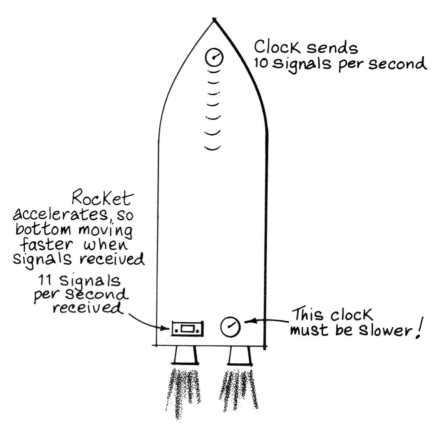

Clock sends
10 signals per second

Rocket
accelerates, so
bottom moving
faster when
signals received

11 signals
per second
received

This clock
must be slower!

Figure 12-4

that *a second at the bottom of the rocket must be different from one at the top!* Since the receiver is picking up 10 percent more signals than the clock sends, *its* second must be 10 percent longer. This means that a clock at the bottom of the rocket actually runs 10 percent *slower* than one at the top. The principle of equivalence assures us that the same thing must happen in a gravitational field.

To see how this helps the astronaut to get his story straight, let us find a formula for the clock slowdown. As indicated above, the clock at the bottom of the rocket must be slower by the ratio of the speed it gains to the speed of light. If the two clocks are separated by a height h, the transit time is h/c. To get the change in velocity, we multiply this time by the acceleration of the rocket,

$$\frac{\Delta v}{c} = \frac{at}{c} = \frac{a(h/c)}{c} = \frac{ah}{c^2}$$

That c squared in the denominator is a pretty big number, so the result is usually tiny. Take a plane flying at a height of 10,000 m. The effect of the earth's 10 m/s^2 gravity is $10 \times 10,000/(3 \times 10^8)^2 = 10^{-12}$. Small as this is, the clocks used by the Naval Observatory are good enough to measure it. To separate it from the effect of the plane's velocity, the clocks were flown on a westbound flight, with the earth's rotation very nearly canceling the motion of the plane. The clocks indeed gained time at higher altitude.

For our astronaut, however, the effect would *not be tiny at all.* In order to reverse a speed of 0.9998c in a reasonable time, his acceleration would have to be immense. In this accelerated frame, earth is a long way "up." Thus during his turnaround, *earth's time would run many times faster than his!*

His stay-at-home twin can now only reply, "I think you're ridiculous to use such a silly reference frame, but I have to admit that your version of the story, bizarre as it may be, is at least self-consistent."

Note that this is *not* a case of two observers each believing the other's clock is slow. Everyone agrees that in an accelerated reference frame the speed of a clock depends on its position and clocks higher "up" run faster. If you have the misfortune to live in such a frame, you obviously must love the clock that's near.

THE FLAT-EARTHERS

Don't be disappointed if you still can't fathom curved space-time. After all, there are still people around who believe the *earth* is flat!

Let two of these benighted individuals begin a journey due north, one starting 100 mi east of the other, traveling at the same speed. We who believe in a round earth know what will happen to them. They will draw inexorably closer to each other, eventually meeting at the North Pole.

If they keep track of their separation along the way, they will find that initially the approach is very gradual, but it speeds up toward the end; *accelerated motion!* If they persist in their flat-earth theory, they will be hard-pressed to explain what went wrong. One way would be to say they were drawn toward one another by a *force!* Just like gravity, their "accelerations" would be independent of their masses, so it would have to be a *force proportional to mass.*

This is just a simplified example, on the two-dimensional curved surface of our planet, of the mistake Einstein claimed Newton had made when he failed to recognize the curvature of four-dimensional space-time. Since this example did not include a time dimension, we substituted the steady progress of our travelers for the familiar steady march of time. Otherwise, the analogy is exact.

Of course this does not necessarily prove that Einstein was smarter than Newton, for he had help. The flat-space geometry of Euclid was the only kind Newton knew about. But nineteenth-century mathematicians had created *noneuclidean* geometries, ready-made for Einstein's use. One mathematician, Bernhard Riemann, even suggested that it might be worthwhile to check the geometry of the space we live in on a large scale, to see if it is really euclidean after all! Let us now have a look at the result of some of these checks.

NEWTON'S LAST STAND

The first test of Einstein's curved-space theory came in 1919, 3 years after he published a prediction of the effect of the sun's gravity on a star image.

We have already seen that the curvature of space near the earth is far too small to have a measurable effect on a light ray. But the sun's gravity is much stronger, at least up close. Fig. 12-5 shows what happens to the image of a star when its light must pass near the sun to reach us.

The curvature of space near the sun is sufficient to bend light through an angle of 1.7 seconds. Small as this angle is, in a photograph taken by a telescope 20 ft long the star's image will be about 0.05 mm from its normal position, which is just within the limit of measurement on good glass-backed plates.

Of course, to photograph stars near the sun you must wait for a total eclipse and take your telescopes to it. The British astronomer Arthur Eddington had to outfit an expedition to the South Atlantic to get his precious pictures. By comparing them with ones taken by the

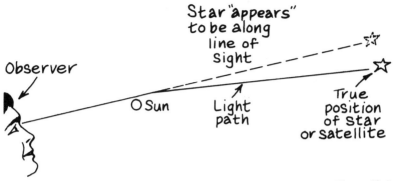

Figure 12-5

same telescope earlier in the year, with the sun out of the picture, Eddington was able to confirm Einstein's prediction.

Today, in the era of radio telescopes, we are far more fortunate, for our sun is a comparatively dim radio star and we need no longer wait for an eclipse. The measurement has been repeated several times in the 1970s, and every new refinement confirms Einstein more precisely.

As soon as he was sure of the result, Eddington sent a cablegram to Berlin to tell Einstein the good news. It was received with Einstein's usual equanimity. When his secretary asked how he would have felt if it had turned out wrong, he replied, "Then I would have felt sorry for the dear Lord, for the theory is correct!"

To be sporting about it, we should really allow Newton one last stand. Couldn't this simply be the result of gravity attracting light? It turns out that this gives a deflection only *half as large as observed,* but we can put a fudge factor in Newton's theory when light is involved and also concede the mysterious effect of gravity on clocks. Can't we then keep our cozy, familiar flat space-time?

But Newton's last stand turns out to be roughly as productive as Custer's. The massacre came at the hands of a little satellite named *Mariner 9,* teamed with interplanetary radar. The Mariner 9 mission of 1972–1973, the first to orbit Mars, carried a very precise radar transponder. This meant that a radar signal from earth could be beamed back immediately. When Mars passed behind the sun, another dimension was added to the situation in Fig. 12-5, that of measuring the transit time of the radar signal.

In flat euclidean space, it is easy to predict the result. Just draw the curved path of the signal on flat paper and measure (or use trigonometry) to get the distance. Then divide by the speed of light. The answer is that the detour around the sun delays the signal by about 3×10^{-8} s compared with the time on a straight path.

But in Einstein's curved space-time, this estimate is as useless as trying to get the exact distance from New York to London off a flat map. The formulas of curved geometry give a delay of 2×10^{-4} s, over *6000 times longer* than the flat-space figure. Irwin Shapiro of MIT headed the team that confirmed the longer delay.

There are other theories of gravity, and Einstein's may not win the last beauty prize. But all the remaining contestants are curved-space theories. *Flat space-time is dead.*

GOODBYE PHYSICS, HELLO GEOMETRY

With the acceptance of curved space-time, the science of mechanics is reduced to utter triviality. Physics disappears, leaving nothing but a geometry. No laws of motion are needed any more, just a single statement:

All objects follow the geodesics of curved space-time.

But what a geometry! That part is anything but trivial. To get just a hint of how complex it is, consider the following. To specify the curvature of a one-dimensional line at some particular point takes but *one* number, the radius of curvature. Since a two-dimensional sheet can be bent in two directions, with any desired angle between them, to specify its curvature takes *three* numbers. Though we cannot visualize what curved three- and four-dimensional spaces are like, we do know how many numbers we need, *six* and *ten*, respectively.

Ten numbers! Newton's theory of gravity got by with just *one,* the force. Clearly, Einstein's is more complex. A whole new mathematical language, *tensor differential geometry,* had to be developed to handle it.

It is so complex that it is useless in most practical situations. Its role in physics is like that of a sacred text, locked away from the multitude but consulted by high priests on sacramental occasions. It is used to verify formulas that apply to very simple situations or approximate ones that give the effects of curved space-time as small corrections to Newton's law of gravity.

Do not expect your intuition to be a reliable guide in a curved four-dimensional world. You would be bucking hundreds of millions of years of evolution of the human central nervous system. Our eyes are our most precious sense organs, and our brains are wired to interpret the information they deliver by the rules of euclidean geometry. The only reason we can grasp the fact that the two-dimensional surface of the earth is curved is because there is a third dimension for it to curve *through.* But it would be a mistake to assume that if four-dimensional space-time is curved, there must be a "fifth dimension" for it to curve through. Those 10 numbers are enough to tell the whole story. We need no outside reference. It's all the space we've got, and it is *fundamentally, indisputably curved.*

THE TAO OF SPACE-TIME

The last link in our new geometric world view is to plug in the formula $E = mc^2$ and see what wonders ensue:

1. Curved space-time means that a field is present.
2. Fields store potential energy.
3. Energy has mass.
4. Mass is the source of gravitational fields; *so back to 1!*

What a long way Faraday's little lines of force have carried us! They started as a way to avoid the problem of action at a distance. Now they generate their own matter. And at the same time, *they are the very fabric of space-time!*

We began with a physics that needed four kinds of reality; *space, time, matter,* and a cause of motion, first *force* and later *energy.* Special relativity forged links between space and time and between matter and energy. Now the unification is complete:

Matter is *energy* is *space-time*

There is beauty enough in these nine syllables to please a Zen master. But while we admire its stark simplicity, let us recognize it for *what it is not* as well as what it is. It is not a completed task but a commitment to a project, a frame in which to hang our picture of the universe. And that picture is *by no means complete,* for gravity is *not* the only field there is.

Electromagnetism is not too much of a problem. There are several ways to build it into our geometry. But the structure of matter is ruled by other fields that operate on the subatomic scale. *We cannot understand the universe until we understand the atom.*

A COSMIC VACUUM CLEANER

There are two neat cases in which the geometry of space-time can be solved exactly. One of these is the universe as a whole, and the other is a large spherical mass, such as our sun or the earth. The latter case leads to one of general relativity's most bizarre predictions, *the black hole.*

Black holes are not really all that new, because they also arise in newtonian gravity. Laplace pointed out as early as 1824 that if a star contains enough mass in a small enough package, the velocity of escape from its surface is greater than that of light. No light can then get out, though light and matter *can fall in.* Simply add the speed limit of *c* from special relativity, and you have a one-way ticket out of the universe; nothing that goes in can ever get out.

Of course in general relativity, unlike Laplace's case, the light does not just *fall back.* It simply travels on *curved paths smaller than the size of the star.* The star is, for all intents and purposes, plucked out of space-time.

The density of matter required is phenomenal. Our sun would have to be only a few miles in diameter to become a black hole. The pressure generated by the nuclear "flame" in its heart prevents it from collapsing. Even when the sun finally exhausts its fuel, we do not expect it to become a black hole but simply collapse to a compact form called a *white dwarf.*

But a star 5 to 10 times heavier than our sun would have gravity enough to pull it down through the white-dwarf stage, through another form known as a *neutron star* or *pulsar* (which is essentially one huge atomic nucleus), to the black-hole stage.

Whether heavy stars actually *do* this is anyone's guess. Stellar collapse usually leads to an explosion, a supernova like the one that launched Tycho's career. The greater part of the star's mass is blown away, and whether enough remains to make a black hole is hard to say. But we do know that enough often remains to form a neutron star; there is one in the center of the *Crab Nebula,* the debris of a supernova recorded by Chinese astronomers in 1054. Since the minimum mass for a black hole is not all *that* much greater than for a neutron star, it is an odds-on bet that they do sometimes form.

For obvious reasons, however, a black hole is well nigh impossible to detect. Our best bet is to catch one that is sucking up matter at a substantial rate. This can happen if the black hole has a nearby binary partner. The hole draws in hot gases from its companion's atmosphere. As they fall, the tremendous acceleration makes the gas radiate light; the higher the acceleration, the greater the frequency. A black hole has strong enough gravity to make *x-rays* come out.

The constellation Cygnus contains a number of x-ray sources. Several of them are known to be binary systems, with *heavy, unseen companions!* From its effect on the motion of its visible partner, the mass of the invisible giant can be calculated. In at least one case, the mass is probably above the upper limit for a neutron star. So as of 1976, quite a few astrophysicists actually believe we are "seeing" a black hole!

Remember what strong gravitational fields do to clocks? As one falls into a black hole, the ratio of its time to that on the outside *actually becomes infinite!* The reader is urged to get a firm grip on something solid while contemplating the implications of that little tidbit. Because this means that as seen from the outside, *the black hole never quite finishes the job of forming!* The last little bit of mass that would push it over the limit *halts at the boundary.* For practical purposes, however, it gets as close to the limit as you want in a matter of thousandths of a second, so the almost black hole behaves just like a fully formed one.

But the point of view of an observer falling into the black hole is the real mind bender. At the boundary, the outside universe *speeds up, flits through its entire lifecycle in an instant, and is snuffed out!* As for the fate of our friend the observer, we must leave it to the imagination of the science-fiction writers.

NOT WITH A WHIMPER

You may find it surprising that *the universe as a whole* is one of the easier problems in space-time geometry, but the explanation is simple: the universe is so huge that the individual stars or even galaxies are of no more consequence than atoms. We can take the distribution of matter as smooth and continuous.

When Einstein solved the problem in 1915, he got a big surprise; *the universe cannot be static. It must either expand or contract.* Since the idea of an eternal, unchanging universe was at that time firmly implanted in the human mind, Einstein did not take the result seriously but added a fudge factor, the so-called *cosmological term,* to make a static universe work. He was later to call this "the biggest mistake of my life," for in 1927 the American astronomer Edwin Hubble demonstrated that the universe *really is* expanding!

What Hubble showed was that all remote galaxies are moving away from our own, as if it were some cosmic untouchable. The farther away they are, the faster they flee.

General relativity tells us that matter and space-time are insepara-ble. The galaxies are *not* simply spreading out in infinite, empty space, but *space itself is finite in size, and it is growing.*

The balloon in Fig. 12-6 illustrates how this can happen. Let the galaxies be spots on the balloon. As it inflates, each spot moves farther from its neighbors. The farther apart they are, the faster they move away.

This was of course a two-dimensional example. Our space has three dimensions. The fourth dimension, time, is the expansion itself. And there being no fifth space dimension, there is no "inside" to our cosmic balloon.

Because of the gravitational attractions of the galaxies, the expan-sion is continually slowing down. Will it eventually go into reverse and become a *contraction?* The answer is that we don't know, because we are

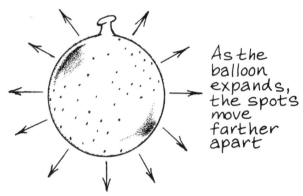

As the balloon expands, the spots move farther apart

Figure 12-6

not sure how much mass there is in the universe. The mass we do see is not enough to turn the trick, by a wide margin. But there may well be more mass out there unseen, waiting to be discovered. Black holes, intergalactic gas and dust, and unknown forms of radiation are all obvious possibilities. Our estimate of the total mass of the universe can only go up.

Projecting the expansion 12 billion years backward in time, we see the birth of our universe in a titanic explosion, called the *big bang* by cosmologists, specialists in the esoteric art of contemplating the origins of absolutely everything.

Anything that rapidly expands must *cool,* and our universe is no exception. It must have been born incredibly hot. Under such conditions, matter conducts electricity freely and is opaque to all forms of light. The universe remained in this state for its first few thousand years. When it became cool enough, light and matter parted company in a brilliant flash that should still be rattling around the universe today. *It has been seen, and looks exactly as predicted by theory.*

Light, too, cools as it expands. What this means is that it shifts to longer wavelengths. Today, this "echo" of the big bang has been stretched into radio waves a few millimeters long. They are known as the 3.2 K background, because their effective temperature is 3.2 kelvins (K), the unit of the absolute (or Kelvin) scale, which starts at absolute zero. The big bang theory is quite explicit about this temperature; how it is measured will be explained in Chap. 15. But the fact that the theory predicts it so precisely is the same sort of "crucial link" that the moon's acceleration was for Newton's gravitational theory.

But how about the *very instant of creation,* when our universe burst forth from what may have been a geometric point? And what will happen if the expansion *does* reverse, and the whole mess rushes back together in one cataclysmic gravitational collapse? About such matters, we can do little more than guess. But our guesses will be far more intelligent after we learn something about the *quantum theory,* the physics of the microworld. So we must now reluctantly drop this fascinating topic, to pick it up again in the final chapter.

THE EINSTEIN CULT

Like Newton before him, Einstein lived to be a legend in his own time. One of the reasons for this, as it was for Newton, was the urgent need of the scientific community for a hero to put on public display. Einstein was ideally cast in the role. With his modesty, his rumpled clothing, his bemused expression, and above all his warm humanity, he was the very antithesis of the cold, dispassionate automaton that had become the popular image of the scientist.

Fame came to Einstein in 1919, in the wake of the eclipse expedition. Eddington deliberately promoted the legend for its political worth. World War I had ripped the international fabric of science asunder. In the wake of the carnage wrought by modern weapons, science and technology stood tarnished with blood. Science needed above all to show the world its humanist face and proclaim anew a tradition that stood above the petty hatreds that set nation against nation.

Eddington proudly announced that British scientists had traveled thousands of miles to prove the theory of a German colleague, in an expedition launched while the two nations were still at war. Furthermore, it was known that Einstein had lived out the war under a cloud because of his pacifist views. The world was ready for a hero like that.

However well Einstein may have suited the role, it most emphatically did not suit him. He had done his best work in obscurity and relished his solitude. The perquisites of celebrity had little appeal to his simple tastes. Near the end of a triumphant world tour in 1921, the year of his Nobel prize, Einstein was conveyed across Spain in the king's own private railway car. He found the experience so distasteful that he completed the journey to Berlin third class.

Liberal social thinkers climbed on the Einstein bandwagon, professing to see in his theory a reflection of their own principles of moral and cultural relativism. How could anyone still believe in absolute values in morality, social customs, or political systems, when not even the perfect world of physics had room for the absolute? Einstein protested: he had been misunderstood. Relativity had not abandoned absolute truth, only the old absolutes of space and time. Relativity had been Poincaré's name for the theory; Einstein would have preferred *Invariantstheorie*.

Even within his own profession, Einstein was to remain a loner, aloof from the mainstream. The better physicists became adept at relativistic calculations, at least with the special theory, but few made the effort to penetrate the deep thought that lay behind it.

This was due in part to Einstein's career, most of which was spent in positions that brought him little or no contact with students. His solitary work habits were another barrier, preventing younger colleagues from mastering his style. And the theories themselves were partly to blame. Special relativity was complete within a few years of 1905. There was no need for a large school of specialists to grind out predictions or for a corps of experimenters to test them. The general theory was even worse, since until recently it was largely inaccessible to experiment.

But most of all, Einstein was isolated by his own personal world view, which swam against the intellectual tides of his day. For Albert Einstein was an uncompromising rationalist, with a deep faith in the underlying logic of the universe. He often personified this faith with the name of God, or "the Old One" (*der Alte*), though he steadfastly refused to affiliate with any organized religion.

In his youth, Einstein had been strongly influenced by the positivist philosopher-physicist Ernst Mach, whose insistence that physics be firmly rooted in a critical examination of the process of observation was an obvious guidepost in the development of relativity. But as he matured, Einstein came to view purely empirical science as "Mach's little horse," a creature which "can only exterminate harmful vermin" but which "cannot give birth to anything living." In this attitude Einstein revealed himself to be a lineal descendant of Galileo.

For in Einstein's universe there was no room for the arbitrary, irreducible fact. There must be no limit to the power of the human mind to reveal why the universe must be exactly as it is. Niels Bohr and his disciples, on whose work we shall soon focus, rejected this view utterly, and even professed to have proved the existence of limits to the power of reason.

But Einstein clung gamely to his lonely credo. In his later years, he was to express it by saying, "What really interests me is whether God had any choice in the creation of the world."

CHAPTER THIRTEEN

The Atom Returns

O NE MIGHT HAVE thought that the birth of relativity was quite enough excitement for one decade. But the opening decade of our century also saw physicists begin to look inside the atom, and what they were to find there proved to be a shock far more unsettling than Einstein's union of space with time. As background to this development, let us begin with the story of how the atom reentered respectable scientific thought during the nineteenth century.

THE ANCIENT ORIGINS OF ATOMISM

It is mandatory to preface any discussion of atoms by paying homage to Democritus, an Ionian philosopher of the fifth century B.C., the earliest known proponent of an atomic theory. Though Democritus' ideas were in many ways strikingly modern and were promulgated by his more celebrated successor Epicurus, his theory never gained wide acceptance in Greek thought. It had largely been forgotten by the time of the late Renaissance rebirth of science. While the dramatic rise of the atomic theory over the last century and a half seems to have vindicated Democritus, only the Greek name *atom* ("indivisible") remains to establish his claim as the father of the theory.

Nonetheless, Democritus' thinking contained the seed of the idea that has dominated twentieth-century physical thought. He was one of the first to perceive that nature on a sufficiently small scale might be qualitatively different in a striking way from the world of our ordinary experience. And he was the first to voice the hope, today almost an obsession, that underlying all the complex richness, texture, and variety of our everyday life might be a level of reality of stark simplicity, with the turmoil we perceive representing only the nearly infinite variety of arrangements of a small number of simple constituents.

Today, the notion that simplicity is to be found by searching nature on a smaller level is embedded in physical thought to the point where few physicists can imagine any other approach.

Despite its failure to gain acceptance, Democritus' theory was able to explain in a qualitatively satisfying way many obvious and simple properties of matter. Its most notable success was the way in which it accounted for the properties of the three *phases* of matter—gas, liquid, and solid. Democritus saw the rigidity of a solid as an indication that its atoms were hooked solidly together. In a liquid they were still close together but free to move around; this accounted for the ability of fluids to settle into the shape of their containers and yet remain as difficult to compress as solids; for the atoms already being essentially in contact, it is hard to press them closer together. It also explained why there is little change in volume when a solid melts into a liquid. A gas, finally, owes its apparently unlimited ability to expand rapidly to fill any container to being composed of widely separated atoms in rapid motion. This picture survives intact to this day, except for a few details that have been filled in.

Democritus' ideas were popular among the philosophically sophisticated founders of modern physics. Galileo, Newton, and most of their contemporaries were atomists, but their beliefs were based more on intuition than on concrete evidence. Moreover, the invention of the calculus had eliminated the difficulties with continuity that had in part motivated the Greek atomists, so the theory received little attention in the century following Newton's work. Still, the atomic theory remained a popular speculation among physicists, because it offered the hope that all the properties of matter might ultimately be explained in terms of the motion of the atoms themselves.

It remained for the chemists of the early nineteenth century to find the first solid empirical support for atomism. Without stretching the point too far, it is fair to say that in 1800 the atomic theory was something physicists believed but couldn't prove, while the chemists were proving it but didn't believe it. Thus, at this point a brief digression into chemistry is in order.

THE BIRTH OF MODERN CHEMISTRY

The latter half of the eighteenth century had been to chemistry what Galileo's time had been for physics. The outstanding achievement had been to put chemistry on a sound, precise quantitative basis. The symbol of the modern chemist of the late 1700s was the balance, which enabled him to substitute precise weights for the crude recipes of the medieval alchemist.

This refined approach helped lead to a number of important discoveries. One of the most significant was clarifying the distinction between a true chemical reaction and a mere process of mixing. This distinction had been dimly perceived before; mixtures displayed properties that were a blend of those of their components, in a manner which depended on their relative proportions. A chemical reaction, however, might produce a substance totally unlike the materials that went into its formation. For example, common water arises from the union of two gases, oxygen and hydrogen. Similarly, the puttylike metal sodium reacts with the green gas chlorine to form ordinary table salt. But at times the basis for the distinction seemed hazy, until the analytic balance revealed the key. By careful weighing one found that mixtures could be formed in any desired proportions, but chemical reactions had an exact recipe. The constituents had to be present in some exact proportion of weights. If too much of one of them was present, some would be left over after the reaction.

Finally, the founders of modern chemistry had clarified yet another distinction between types of substances. They classified some as *elements,* which could not be broken down into other substances, and others as *compounds,* which could.

The whole picture was terribly inviting to an atomist. All one need do is identify the elements as representing the different kinds of atoms, chemical compounds as substances formed by attaching atoms of different elements together, and mixtures as a free mingling of independent atoms without any ties between them. But atomism, and indeed the whole intellectual style of imaginary model building that lay behind it, was mainly the province of physicists. One such, the Italian Amedeo Avogadro, pushed the atomic idea in chemistry very forcefully well before the end of the eighteenth century. But while his arguments satisfied many of his own colleagues, the vast majority of chemists remained skeptical of such wild talk. Throughout its history chemistry has tended to be a far more conservative science than physics, sticking close to its empirical roots and disdaining abstractions and speculations. The chemists paid little attention to atomism until one of their own number, the English chemist Thomas Dalton, brought it forcefully to their attention by showing that an atomic structure to matter could explain the peculiar regularities that kept popping up in the recipes uncovered by the analytic balance.

This regularity was expressed in the *law of constant proportions.* Stated crudely, it indicated that the amounts of an element that entered into forming all its compounds were related. Hydrogen, for example, was always vastly outweighed by its partner when entering into combination, while lead always dominated its compounds.

In more exact terms, it was found that each element seemed to have a characteristic *equivalent weight*. Hydrogen was the lightest and could be taken as the starting point on the scale. Oxygen was 8 times heavier, sodium 23 times, chlorine 35, and so on. All recipes for compounds could be formed from these equivalent weights.

In the first decade of the nineteenth century, Dalton pointed out that the whole scheme could be simply understood by taking the equivalent weights to represent the relative weights of the atoms of the elements. Then the recipe for common salt, 23 parts sodium to 35 parts chlorine, merely represented the fact that chlorine atoms were 35/23 as heavy as sodium atoms, and salt was formed by joining each sodium atom to one chlorine atom. Such a combination Dalton christened a *molecule*. The molecule is the smallest constituent of a chemical compound, just as an atom represents the smallest unit of an element.

Still, a lot of facts remained unexplained. Some elements seemed to have more than one characteristic weight. It gradually became clear that the simple pairing off of elements into two-atom molecules proposed by Dalton was too simple; some molecules must contain three or more atoms. For example, oxygen atoms prove to be 16, not 8 times as heavy as hydrogen. The proportion 8:1 of oxygen to hydrogen in the recipe for water reflects the fact that two hydrogen atoms join each oxygen atom when a molecule of water is formed.

The whole situation was very confusing. As is usual in such situations, there was of course a certain amount of bad data in circulation to confuse the issue further. It took 50 years to untangle the mass of chemical data, but in 1858 the patient Italian chemist Cannizzaro published a book that finally established the correct relative weights of the atoms of the better-known elements and gave the atomic composition of their known compounds. The atomic theory has been the foundation stone of chemistry ever since.

THE PHYSICISTS PICK UP THE BALL

The success of atomism in chemistry was bound to encourage the physicists in their natural predilection for the theory. Old ideas were resurrected, cloaked in a new mantle of respectability. One important idea dated from Newton's celebrated contemporary (and rival) Robert Hooke, whose claim to have independently discovered the inverse-square character of gravitation had touched off one of the first priority fights in the history of physics. A confirmed atomist, Hooke speculated that the outward pressure exerted by a gas on the walls of its container might originate in a hail of tiny atoms. Each atom exerts a force on the wall when it hits, and there are so many such impacts that the result seems a constant outward push.

Hooke found support for his view in the careful penumatic experiments of Robert Boyle, performed a generation before Newton and Hooke. Boyle found empirically that if a gas is compressed in a closed container, as shown in Fig. 13-1, the pressure on the walls varies inversely with the volume available to a gas. For example, if the piston is pushed in far enough to reduce the volume to half its original value, the pressure of the gas is thereby doubled.

This effect is quite easy to understand in atomic terms. With half the volume to roam in, the atoms are packed in closer, so there are twice as many of them in any region of the cylinder. Since their motion presumably is unaffected by this crowding, they strike the walls with the same impact, but there are twice as many impacts taking place, thereby doubling the pressure. This alone was a rather weak boost for Hooke's atomic theory, but later experiments on the effects of heat on gases gave it firmer support.

Even before the days of Galileo the notion that heat might represent some form of microscopic motion enjoyed some vogue. Francis Bacon, the fifteenth-century English philosopher, subscribed to the theory. After the discovery of the mechanical energy equivalent of heat, as a consequence of the work of Count Rumford and others, the idea became even more appealing. But far more compelling evidence could be found in the very regular behavior of gases when heated or cooled. Heating a gas in a sealed container always causes a rise in pressure. If heat were somehow related to atomic motion, this seemed natural, but again the speculation lacked the sort of quantitative handle required for a convincing test. The issue was further clouded by the

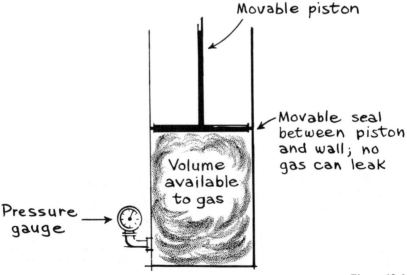

Figure 13-1

arbitrary character of temperature measurement. Zero on a thermometer, whether Fahrenheit or Celsius (centigrade), doesn't represent zero *of* anything: it is merely an arbitrarily chosen point, a convenient cold temperature. Gabriel Fahrenheit had chosen the lowest temperature he could produce in his laboratory, the freezing point of very salty water, as the "zero" for the scale now used only in the United States. Celsius, whose scale is in wider use, chose the freezing point of fresh water.

Studies on the behavior of gases heated in closed containers showed that the pressure seemed to rise linearly with temperature, as shown in Fig. 13-2, but zero on whatever thermometer used seemed to have no special significance. There was still some pressure from the gas at zero degrees, and if one cooled below zero, the pressure continued to drop linearly, as long as one stayed well above the boiling point.

Yet by Dalton's time, careful work on the temperature-pressure relation in gases revealed a significant clue. No matter *what* gas was used or what might be its original pressure or the shape of the container, the pressure always behaved as if it would reach zero at the same temperature, −273°C. This is also illustrated in Fig. 13-2. If gases under different conditions were studied, for example, by starting with compressed gas like the air in an ordinary tire, the pressure would be higher but would also rise and fall faster with changes in temperature. The graphs always crossed zero pressure at −273°. Gleeful atomists christened this temperature "absolute zero" and hypothesized that it represented the cessation of all atomic motion, for that would bring an end to the pressure by ending the hail of molecules on the walls.

Though the atomic case was building, its empirical props remained weak. Absolute zero itself was rather hypothetical. No means of cooling was available to let one get anywhere near it; even as refrigerators improved, most gases annoyingly dropped out of the game by going liquid, somewhere above absolute zero.

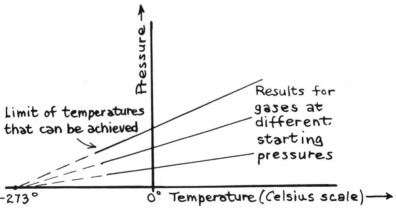

Figure 13-2 Pressure changes due to heating and cooling gases.

But in 1847, Rudolf Clausius showed that one could account completely for the behavior of gases by one simple hypothesis: *the "absolute" temperature of a gas is merely a measure of the kinetic energy of its molecules.*

To demonstrate this required little in the way of mathematical sophistication. If pressure is due, as Hooke guessed, to a hail of molecules, speeding them up raises the pressure in two ways. First, the rate of collisions with the walls of the container increases. This effect would of course be proportional to the speed of the molecules; if a molecule moves twice as fast, it will strike the walls twice as often. Second, the force due to each molecular impact is increased. The *force* exerted by a molecule on impact is, by Newton's second law, proportional to the change in its *momentum*. The two effects taken together indicate that speeding up molecules will cause the pressure to increase proportionally to the product of the momentum and the speed:

$$\text{Pressure} \propto (mv)v = mv^2$$

and thus in proportion to the kinetic energy. Thus, equating absolute temperature with kinetic energy of the molecules would account for the pressure-temperature law as well as Boyle's law.

Clausius' theory also gave a natural explanation of the role of heat in the law of energy conservation (see Chap. 5). If temperature is merely a measure of the kinetic energy of molecules, the conversion of mechanical energy into heat merely represents the conversion of the motion of a large object into the random motion of its individual atoms. No real conversion has taken place—kinetic energy resulting from a combined motion of a large number of atoms in the same direction has merely been changed into kinetic energy of motion in different directions, and since energy is a nondirectional measure of motion, this does not matter.

Note that we have said *average* kinetic energy. It readily became apparent to the atomists that it was unreasonable to assume that all molecules in a gas were moving at the same speed. Even if at some instant they were, the collisions between molecules would quickly disrupt this order. For example, a "sideswipe" collision between two molecules moving at right angles to one another could stop one dead, by transmitting its forward motion to the other, as shown in Fig. 13-3. But this presented no difficulties for the theory; the pressure being generated by a fantastic multitude of impacts, the slow ones make up for the fast ones.

The details of the picture were filled in over the next few decades after Clausius by a number of gifted contributors, most notable among them the same James Clerk Maxwell who unified electromagnetism. This complete picture of gases as composed of widely separated, rapidly moving molecules is known as the *kinetic theory of gases*, the success of which turned most physicists and chemists into convinced partisans of

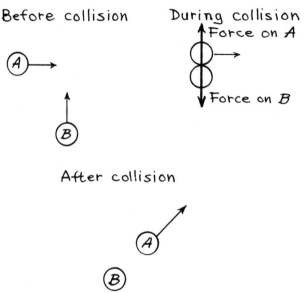

Figure 13-3 How collisions change speeds of atoms.

the atomic theory. By the end of the nineteenth century the picture was complete.

HOW BIG IS AN ATOM?

One vexing problem remained to cloud the success of the atomic theory; neither physicists with their gas laws nor chemists with their reaction recipes had any clue to how big their atoms were or any way to tell how many there were in a given substance. The same pressure could be obtained in a gas by having more molecules each individually lighter. Put another way, the chemists could establish that an oxygen atom was 16 times heavier than one of hydrogen, but no one knew how much either actually weighed.

Interestingly enough, the long-sought direct evidence for the atomic theory and a means of estimating the size of atoms had been at hand, yet overlooked, since 1827, in the form of an accidental discovery by the botanist Robert Brown. Observing through a microscope pollen grains suspended in water, he was disturbed by a phenomenon that had plagued microscopists looking at exceptionally small objects for over a century: the pollen grains refused to sit still, but instead insisted on hopping about in a jerky erratic fashion. Convinced in advance that pollen was an inert spore possessed of no means of locomotion, Brown proved his point by showing that the motion was most rapid for the smallest grains, hardly the sort of behavior expected out of a living microorganism. But how natural for the atomists! Finally they had found an object small enough to be disturbed by the unevenness of the hail of molecules, one on which a few particularly swift molecular impacts that happened to come on the same side could pile up to produce a noticeable effect. But the nineteenth century was nearly over and the controversy over the existence of atoms in full swing before the physicists rediscovered Brown's work. And not until 1905 did Einstein, in his fabulously productive year, work out the quantitative details to permit an estimate of absolute atomic weights from measurements of Brownian motion. He used this work, rather than the more controversial ones, as his Ph.D. thesis.

From the fact that in liquid and solid elements the density is more or less proportional to the atomic weight, it was clear that atoms are all pretty much the same *size*, despite the disparity in weights. Using a variety of evidence, by the turn of the century it began to appear plausible that typical atoms are about 10^{-8} cm in diameter, and that there are about 6×10^{23} atomic weight units in a gram. Thus, as physics entered the twentieth century, the atom was no longer regarded as a hypothetical and probably unobservable object but a fit object of study in itself, if only the right probes could be found to look into this tiny world beyond the range of the finest microscopes.

WHAT'S "INSIDE" AN ATOM?

The natural next step was for a few bold souls to speculate on what an atom might really look like. Though atomism in its naïve infancy presumed atoms to be simple and nearly structureless, by 1900 there was plenty of evidence to the contrary. First and foremost, some means had to be found for hooking atoms together into molecules. Secondly, the existence of trends and similarities in chemical properties indicated by the periodic table of the elements was strongly suggestive of underlying structure. Still, the word *atom* meant *indivisible,* and it was not respectable to speculate much about dividing the indivisible until two sensational discoveries just before the turn of the century made it obvious that atoms were by no means structureless and furthermore that there were experimental means to take them apart. These were the discovery of radioactivity by Becquerel (1889) and that of the electron by J. J. Thomson 7 years later.

Becquerel's discovery had the more sensational popular impact. One of the marvels of the eighties had been the x-ray. The practical implications of a form of "light" that could penetrate opaque objects titillated the late Victorian public and led to sensationalized and amusing newspaper articles. Following a hunch, Becquerel tried to find a substance that would give off x-rays when placed in ordinary light. Instead, he found that minerals containing the element uranium did indeed give off penetrating radiation, but it was not identical to x-rays and seemed to arise spontaneously, not only in the absence of light, but in fact oblivious to all outside influences. No amount of heating, treating, or cajoling could change the inborn rate at which a radioactive substance gave off rays. What was even more sensational, within a few years Ernest Rutherford (of whom we shall hear a great deal more in the next chapter) established that the emission of radiation was accompanied by a chemical change—one that transformed one element into another!

It is hard to imagine any discovery that could have been more shocking. The proudest boast of nineteenth-century chemistry was that it had proved the futility of the medieval alchemist's search for a means of turning base metals into gold. The atoms were immutable, and there was no way to *produce* an element; one simply had to go out and find it. Despite the existence of radioactive disintegration, a spontaneous process, there still remained no means of transmuting the elements by means of a chemical reaction, and thus the basic ideas of chemistry were not threatened. If an atom occasionally decided to change its identity, and there was no way to persuade or dissuade it from this step, chemistry was little affected. But to physicists, the impact of the discovery of radioactivity was to suddenly make probing within the atom not only respectable but imperative. Anything that could up and change in this fashion must have internal workings of quite a complex order.

The leading clue to the structure of the atom was the discovery of the electron, found in studies of electric currents in gases, a phenomenon familiar to anyone who has seen a neon sign.

IONS AND CATHODE RAYS

Back in the 1830s Michael Faraday had done studies of the conduction of electricity in liquids. There the flow of electricity is usually accompanied by an actual movement of material to the points where the current entered or left the liquid. For example, the passage of electric current through water results in the liberation of hydrogen at one electrode and oxygen at the other, a phenomenon known as *electrolysis*. Faraday found that the amount of an element that arrived at a point of electrical contact was *proportional to the total electric charge flow and to the chemical equivalent weight of the element,* the same equivalent weight discovered from the law of constant proportions. To anyone who believed in an atomic picture of matter, the interpretation of this law was obvious: all one needed was that electricity be transported in the form of an electric charge on the atoms. If for some unknown reason all atoms regardless of type carried the same electric charge, the total material flow corresponding to a given electric flow would depend only on the atomic weight, explaining Faraday's law. The charged atoms were christened *ions,* a Greek word meaning "wanderer" and thus a fit companion for the term *atom* itself.

In later Victorian England, J. J. Thomson was the most celebrated experimenter in a proud imperial nation boastful of her scientific achievements, and in his study of electrical conduction in gases, important to the mushrooming electrical industry, he was working on a fashionable problem. In the spirit of Faraday's work with liquids, many experimenters sought to identify the carriers of electricity in gases. They reasoned that if the gas were sufficiently rarefied, its atoms could travel great distances without colliding with one another. In unimpeded flight, ions in the gas could be studied by applying electric and magnetic forces from outside the tube and observing the deflection.

Though still no one knew the absolute mass of an atom, the amount of charge carried per unit mass of ions had been established by Faraday. This would be sufficient to predict their response to electric and magnetic forces, for these forces are proportional to the electric charge. Thus, a larger mass would be offset by a proportionally larger force because of the larger charge; the acceleration produced by an electromagnetic force is thus determined solely by the ratio of charge to mass.

The negative charge carriers were the first ones found. If a hole is pierced through the positive electrode in a gas tube, the negative carriers fly through and leave a telltale glowing spot where they touch the glass. Such an experimental setup is shown schematically in Fig. 13-4.

But the behavior of these *cathode rays,* as they came to be called, seemed at odds with the notion that they were like Faraday's ions. For one thing, they had identical properties regardless of the gas used to fill the tube. Secondly, they easily penetrated very thin sheets of matter placed in their path. If they resembled atoms, they should rebound from the densely packed atoms of a solid material. Thus, they might represent

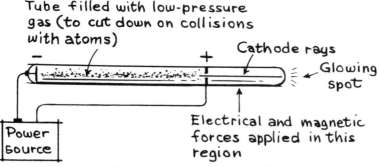

Figure 13-4 Apparatus for the discovery of the electron.

a new form of matter, possibly one smaller than atoms and therefore more able to penetrate.

Finally, while moderate magnetic fields easily deflected the beam of cathode rays, it took enormous electric fields to budge them from their paths. To Thomson, this was the key to the mystery. It suggested that they were moving very fast. For the magnetic force on a charged particle is proportional not only to its electric charge but also to its speed, while electric force depends on the charge alone. Thus, magnetic fields are far more effective in deflecting fast-moving particles than electrical fields, which do not benefit from the increased velocity.

In fact, since all other factors are equal, comparing the relative strengths of the electrical and magnetic fields required to produce equal deflections of the cathode ray particles enable Thomson to compute their velocity. They proved to be fast, indeed—more than one-tenth the speed of light, an unprecedented speed for material objects!

Once the speed was known, Thomson could use the deflection to calculate the acceleration, using the formula in Chap. 3 for acceleration in curved motion. From the acceleration he got the ratio of charge to mass. The result was surprising: nearly 2000 times higher than the charge/mass ratio of hydrogen, the lightest known ion. Since the ratio proved independent of the gas in the tube, Thomson in 1897 announced the discovery of a new form of matter, which was christened the *electron*.

Either Thomson's electrons carried an enormous charge compared with ions, or they were far lighter. All the evidence as to their speed and penetrating power pointed to the latter conclusion: electrons, if they carried the same charge as ions, were 2000 times lighter than a hydrogen ion.

Shortly afterward, studies of the positive carriers moving opposite to the electrons in the tube confirmed that they were ordinary ions. Since the electrons apparently arose from the atoms of the gas, it seemed reasonable to presume that they carried the same charge as the positive ions, to make the electrical charges balance when the electrons were recombined with the positive ions.

THE ELECTRON AND THE ATOM

Thomson's electron was immediately hailed by some as the "atom of electricity," the solution to the mystery of the nature of electricity. Of course, its positive counterpart eluded discovery, but the existence of this unsuspected light object seemed to explain the great mobility of electricity. But the true significance of his discovery was by no means lost on Thomson; since it had been produced from the dissociation of gas atoms, his electron must be a component of the atom itself. Where lesser minds saw the solution to an old mystery, Thomson saw the opening of a new adventure. A potential building block of the atom had been uncovered, and the rush was on; led by Thomson himself, the more imaginative citizens of the world of physics began inventing hypothetical models of the atom, not as a mere pastime or for purposes of illustration, as a few years earlier, but in dead earnest. A few clever experiments might tell physicists what really went on in an atom!

In their bold optimism, these early speculators on atomic structure little realized that they had opened a Pandora's box and that the demons resident therein had the power to bring down the edifice of classical physics patiently erected over three centuries.

CHAPTER FOURTEEN

Rutherford Picks Apart the Atom

---

*I*MAGINE A GROUP of proud and inventive people quarreling over the contents of a locked box and you have a pretty good picture of the mood of the early thinking on atomic structure. It only took a few years for most of the protagonists to sift out into one of two camps: the *planetary* and the *plum-pudding* schools.

Given the similarity between Coulomb's law of electrical force and Newton's law of gravitation, an atom that resembled the solar system with an electrically positive sun and negative electrons for planets was too pretty an analogy to pass up. Furthermore, it placed at the disposal of its supporters the powerful computational tools developed over two centuries of study of the motion of the planets.

But the opposing camp had at *its* disposal the formidable authority of its founder, none other than the illustrious J. J. Thomson himself, soon to become one of the first Nobel laureates in physics for his discovery of the electron. He proposed a sphere of positive charge in which the electrons were embedded, as shown in Fig. 14-1; the descriptive albeit pejorative term "plum pudding" was his own choice.

LIGHT EMISSION IS THE ARBITER

The debate would have been abandoned as sterile had there not been a body of data crying for explanation by an electrical model of the atom. These were the data on the emission of light by atoms.

Light from a source containing a single element in a gaseous state,

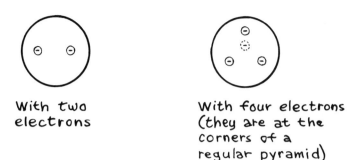

With two
electrons

With four electrons
(they are at the
corners of a
regular pyramid)

Figure 14-1   Plum-pudding atoms.

such as a neon sign or the mercury lamps now widely used for street lights, always has a characteristic color. When this light is broken up into its component colors by a prism, a striking result is obtained. Instead of a continuous spectrum (rainbow), which occurs when a solid or liquid is heated to glowing, one finds the light is composed of a few very pure, sharply defined colors. The best way to observe this is to have the light originate from a thin slit, as illustrated in Fig. 14-2. If it then falls on a viewing screen or photographic plate, the result is a series of thin lines, each of a different color. For this reason, this type of light pattern is referred to as a *line spectrum*. For most elements, only a few lines are bright enough to be seen with the naked eye. But if a long photographic exposure is made, many fainter lines appear; for some elements, hundreds have been catalogued.

Ever since Maxwell uncovered the electromagnetic nature of light, it had been clear that there was only one way to produce light: somewhere, an electrical charge must be going through a regular, periodic motion. The frequency of this motion determines the frequency of the light; and since light regardless of color travels at the same

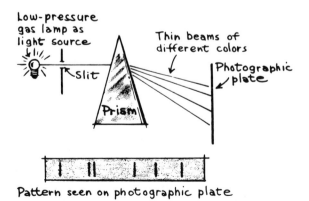

Pattern seen on photographic plate

Figure 14-2 Apparatus for producing line spectra.

speed, the frequency determines the wavelength. This was one reason why Thomson's insistence that his electrons must be part of ordinary matter was so readily accepted; Maxwell's theory implied that anything that emitted light had to be electrical in nature.

It was also clear that in a gas the light must be emitted by individual atoms. Not only are the atoms separated by many times their own size, but the oscillations of light waves have enormous frequencies, around 10^{14} Hz. There are many oscillations in the time between encounters with another atom, and thus it is hard to imagine how many atoms could cooperate to produce such a rapid oscillation. Furthermore, the striking differences between the continuous spectrum of light emitted by densely packed solids or liquids and the line spectrum from rarefied gases gave support to the notion that line spectra represent light from individual atoms.

Thomson's model gave a quite natural explanation for this light emission. He imagined his electron plums were able to move freely in their positively charged pudding, held in place by a delicate balance between their attraction to the center of the positive charge and their mutual repulsion. He and his supporters devoted a great deal of energy to finding the patterns the electrons would assume. If disturbed from this pattern by a collision between atoms, they would oscillate around their normal positions, just as a pendulum oscillates when disturbed from its equilibrium point of hanging straight down. Whenever a charge is accelerated, as indicated in Chap. 6, its electromagnetic field changes. This change radiates out in the form of light, which carries away the energy of motion, so that the vibration would gradually die out like the motion of a pendulum dying as it yields up its energy to air resistance. It took no great skill to calculate the natural frequency of these oscillations, which would set the frequency of the waves emitted; for atoms around the known atomic diameter of 10^{-8} cm, they were appropriate frequencies for visible light. This was a very encouraging result: that the electron had just the right amount of charge and mass to connect the 10^{-8} cm of atomic size to the 10^{14} Hz of light frequencies had to be more than a coincidence. The plum-pudding atomists were sure they must be on the right track.

The same numerical coincidence also encouraged the planetary enthusiasts. Orbits around 10^{-8} cm in diameter gave the right frequency of rotation to produce visible light. But this also proved the undoing of the model. Unlike Thomson's atom, there was no natural way to stop the light emission or to give it a definite natural frequency. An orbit could have any size and thus could radiate light at any frequency. As the electron lost energy, its orbit would gradually shrink. Spiraling in to its doom, the electron would gradually change its frequency of rotation, unlike Thomson's electron oscillations, which kept the same frequency as their amplitude decreased. Even worse, simple calculations using

Maxwell's laws showed that an electron took no longer than a millionth of a second to complete the spiral into the nucleus. Thus, the planetary atom was unstable and gave no natural explanation of the line spectrum. Despite heroic and ingenious efforts to eliminate these faults, the model fell into disfavor.

Still, until the Thomson model could be shown to explain the observed spectrum lines in all their quantitative detail, the field was open to all comers. And the quantitative detail was immense. The wavelengths of spectral lines are among the easiest physical quantities to measure to high precision; 1 part in 100,000 is not uncommon. Furthermore, precise data had been piling up for decades, thanks in part to their practical value. The set of spectrographic lines produced by an element is its fingerprint, and spectrographic analysis was a marvelous tool for chemistry. One bright line could reveal a small trace of one element in the presence of another that had only faint lines near the same wavelength.

While the chemists were content to amass their fingerprint file, physicists were doing much of the actual experimental work, because the techniques involved fell in the realm of optics. And the physicists could not resist the temptation to search for regularities in their data.

ORDER IN THE CONFUSION OF SPECTRAL LINES

With the added spur of the debate on the structure of the atom, the search for order in the confusing mass of data on spectra hit a peak around the turn of the century. By the first decade of the new century, two striking features had emerged.

The simpler of the two, but the more difficult to establish because of the immsense amount of data that had to be examined, was the *combination principle* enunciated by W. Ritz. He observed that, far too often to be explained by coincidence, adding the frequencies of two lines in the spectrum of some element gave the frequency of a third. In the case of elements where the fainter lines had been carefully mapped out, nearly all the lines fit into such a relationship. Allowing that to cover the exceptions there might be a few lines too faint to have yet been discovered, Ritz formulated a general principle: *all spectral lines have frequencies that are either the sum or the difference of the frequencies of two others.*

A portion of the line spectrum of helium. (Mount Wilson and Palomar Observatories.)

There was no simple way to account for this principle in Thomson's atomic model.

The second important regularity applied only to the spectrum of the element hydrogen. As early as the 1880s, it had been recognized that hydrogen, the lightest of the elements, also had a far more orderly spectrum than the others. A Swiss schoolteacher by the name of Balmer found that the frequency of every line in the hydrogen spectrum can be obtained from a single formula:

$$\nu = \text{const} \times \left(\frac{1}{n^2} - \frac{1}{m^2}\right)$$

where n and m are integers (whole numbers). Later, the Swedish spectroscopist Rydberg showed that the same formula very nearly held true for the elements that most resemble hydrogen chemically, the alkalis that share with hydrogen the first column of the periodic table. Once again, the Thomson model offered no simple explanation of this formula.

But while the model builders were struggling with the problem of how to impose these regularities on light emitted by a plum-pudding atom, a surprise experimental result from the Manchester laboratory of Ernest Rutherford indicated they were betting on the wrong horse. Rutherford was such a towering figure in the physics of his time that it is fitting to pause here and introduce the man.

ERNEST RUTHERFORD

In the 1920s, when he was at the zenith of his fame and recently elevated to the peerage as Lord Rutherford of Nelson, Ernest Rutherford was told by an envious colleague that he was "lucky to be riding the crest of a wave." Rutherford retorted, "Lucky, nothing!—I *made* the wave." While this rebuke is hardly indicative of excessive modesty, it was perfectly justified. Practically everything known about radioactivity and the atomic nucleus, and by then a great deal indeed was known, was the work of Rutherford, his students, or his coworkers. His fierce pride never allowed him to forget the few important discoveries in his chosen field that did not bear his name. Not since Faraday had one man so dominated an experimental discipline. There was no way to beat Rutherford—you just had to join him. Young physicists flocked to his laboratory, and in an environment where several startling discoveries per year were regarded as commonplace, they stretched and developed their talents to the full. It wasn't even safe to stay out of his field; if an exciting problem arose in any area of physics, Rutherford might very

Rutherford (right) with Geiger in the Manchester Laboratory.

well pounce on it. And the more important it was, the more likely that he would beat the specialists in that field at their own game.

Rutherford's fame began almost the moment he stepped off the boat in England in 1895, a raw colonial from New Zealand. Some clever studies on the effect of radio waves on magnetic iron, conducted under unbelievably primitive conditions in a converted coat closet at Canterbury College in Christchurch had earned the ambitious 24-year-old a scholarship to Cambridge as a research student at J. J. Thomson's Cavendish Laboratory. The Cavendish was the most outstanding example of a new genre of university research laboratory that had developed with the secularization of universities in the nineteenth century.

The dominant activity at universities since their creation in the Middle Ages had been the training of the clergy, with as secondary duties laying the groundwork for such learned secular callings as medicine and law. Within this framework, experimental work in the sciences was almost an outside activity, a private activity of the professor in which the university as an institution could hardly be expected to take much interest. But with the growth of the natural sciences, a pattern emerged in which the university itself, with the aid of funds from private donors, provided the professor with a laboratory. Working under his direction in this laboratory would be a number of more junior researchers who had not yet reached the olympian level of professor, a title still reserved to a small elite even today at most European institutions. Normally there was only one professorial chair in each field. At the top of the scale, the professor's assistants included established scientists awaiting his retirement or demise, that of a professor at another institution, or creation of a new chair at a new or growing school. Below this came various positions of varying degrees of permanency, down to the lowly research student, usually but not necessarily a candidate for the doctoral degree.

This system did much to encourage the internationalization of science and the free flow of ideas. Under it, a young scientist with high ambitions was more or less forced to move every few years. Any one nation of course had a limited number of laboratories in his field. By the time he settled down, he was likely to have had some degree of personal contact with most of the people working in his area, at home and abroad.

J. J. Thomson was among those professors who permitted a great deal of independence to the junior staff, and in this free environment Rutherford quickly made his mark. In only three years he had attracted enough attention to be offered a chair at McGill University in Montreal, Canada. Though a native-born Englishman might have regarded this as an exile with a dubious future, as a colonial Rutherford had few qualms about taking his chances at a fast-rising institution that was already one of the best in the Empire outside the mother country. And despite Rutherford's youth, his own reputation and the praise of Thomson were sufficient to attract a stream of first-class assistants from Britain, the United States, and the Continent as well as Canada itself. The productivity of his laboratory over his 9 years in Montreal was phenomenal, and in 1907 he returned to England to assume a chair at the University of Manchester. Later he was to cap his career by succeeding Thomson himself as the director of the Cavendish Laboratory, but it is in his Manchester days that we find the most of interest for our current topic, for it was there that he administered the *coup de grâce* to the plum-pudding atom.

THE CANNONBALL IN THE HAILSTORM

Observing what happened to the radiation emanating from radioactive substances when it passed through matter had been just part of Rutherford's bag of tricks at McGill, where his primary task had been to identify the composition of the radiation. This problem solved (and the 1908 Nobel prize earned thereby), Rutherford had the insight to guess that the technique might be turned around, that the now well-understood radioactive emanations could be used as a probe to see what might be inside matter. Nagging his thoughts was the observation that alpha radiation, which he had shown to consist of helium atoms with two electrons stripped away, was deflected somewhat when passing through thin sheets of mica. It was a small effect which most experimenters might well have overlooked. But Rutherford quickly realized that his alpha particles were too heavy and too fast-moving to be budged from their paths except by a strong electrical force. Careful measurements of the deflection could be used to reason back to the size of the force, which might in turn give a clue to how electrons were arranged in an atom.

But like most of his contemporaries, Rutherford had little doubt that Thomson had found the right picture of the atom. A mere corroborative experiment, and one that might prove difficult to interpret, was hardly worth his personal attention. However, there was a young research assistant named Marsden available for the task. A check on whether there might be anything interesting in alpha scattering would make an ideal assignment for Marsden to cut his teeth on.

The prospects for anything interesting coming out of the experiment were not very bright. An alpha particle approaching one of Thomson's plum puddings, as shown in Fig. 14-3, would experience no force until it got very close to or inside the atom, for the negative electrons would balance the positive charge on the pudding. Once inside, the forces would be considerable, but they would be exerted mainly by the electrons. Since these were many times lighter than the alpha particle, the electrons rather than the heavy, swift projectile would be the most disrupted by the encounter. It would be like a cannonball fired into a hailstorm. After traversing many atoms, the cumulative

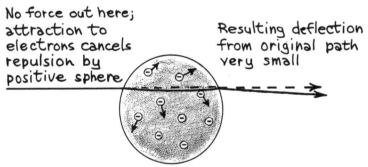

Figure 14-3 Passage of an alpha particle through a plum-pudding atom.

effect of many small encounters with electrons might have deflected the alpha particle a bit, but no large deflections could be expected. If the deflections resulted from many small scatters, it seemed unlikely that much detailed information about the structure of the atom would be retained.

So Marsden seemingly had little to look forward to, even considering that he was a novice research student inured to routine laboratory jobs. The task he faced is depicted in Fig. 14-4. Inside a vacuum chamber (to prevent atoms in the air from interfering), he had to place a thin tube containing a source of alpha radiation. This produced a narrow beam of alphas emerging from the tube. The beam had to be narrow in order to permit observation of tiny deflections as the alphas traversed the target, a thin sheet of gold leaf. Gold had been chosen because since medieval times craftsmen had mastered the art of hammering this soft metal to an incredible thinness; good gold foil is translucent. This was essential, because even a sheet of cardboard is sufficient to stop a beam of alphas (it is the more penetrating beta and gamma particles that are primarily responsible for the fearsome reputation of radiation).

To detect the alphas, Marsden had to patiently count the tiny flashes produced when they struck a fluorescent screen; to see these flashes reliably required long hours of adaptation to a dark room.

On the whole, the results were not terribly surprising. On the average, the alphas were deflected by only a few degrees. But a very few of them, perhaps one in a thousand, were deflected through substantial angles. Some even came off backwards!

Again, many physicists might have been content that the average scattering was reasonable, but Rutherford was nagged by those few scatterings through large angles. If they resulted from a cumulative effect of many small scatters, we would expect the average scatter to be

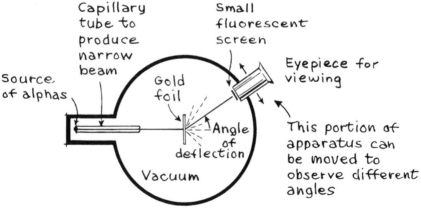

Figure 14-4 Geiger-Marsden apparatus.

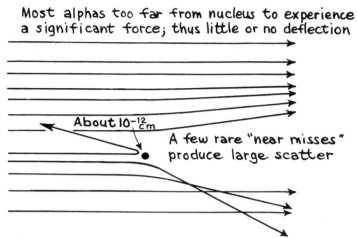

Figure 14-5 Scattering of alphas by a nucleus.

larger. Marsden's immediate research supervisor, Hans Geiger (inventor of the Geiger counter), got directly involved in the work, and a systematic quantitative study was undertaken.

By 1911, though the data were still crude, Rutherford was sure the results ruled out the plum-pudding atom. Instead, he suggested that the positive charge on the atom might be confined to a tiny region; the large-angle scatters came from single close encounters with this *nucleus*, as he called it.

With this picture of the atom, shown in Fig. 14-5, it is easy to explain both the small average scattering angle and the occasional large one. With the atom mostly empty space, the alphas rarely came near a nucleus. Those that did would experience tremendous forces, since Coulomb's law gives a force that varies inversely as the square of the distance. Since the nucleus contains most of the mass of the atom, the cannonball is meeting a *bigger* cannonball off which it can recoil backward. So the small average deflection comes from the fact that most of the alphas traverse the gold foil without ever getting near a nucleus. Yet the few large deflections can be explained as examples of the rare near misses.

More importantly, it was possible to exactly calculate the pattern of this type of scattering. In the Thomson model, the scattering resulted from many tiny deflections from encounters with individual electrons. The result would thus be a pattern that followed the familiar bell-shaped curve of random processes. But in Rutherford's picture the paths of the alphas could be calculated exactly from Newton's and Coulomb's laws, for there is nothing simpler than the motion of a body repelled from a single center of force. The result could be expressed in a simple mathematical statement: the number of alphas per minute hitting the fluorescent screen at some particular angle was inversely proportional to the sine of one-half of the angle, raised to the fourth power.

Geiger and Marsden pushed on with ever more careful measurements, in order to check this quantitative prediction in detail. In 1913 their final results were published. The data fit an inverse-sine-to-the-fourth curve beautifully. Table 14-1 gives the actual data as they appeared in their article in *Philosophical Magazine* (vol. 25, p. 604), a physics journal whose name betrays its ancient origins, when physics was still "natural philosophy." It is the third column of this table that spelled the death of the Thomson atom. Though 4000 times as many flashes are seen on the screen at 15° as when it is placed at 150°, dividing the measured numbers of the second column by the computed ones in the first column, which give the value of $1/(\sin \frac{1}{2}\theta)^4$, results in a number that is nearly the same for all measurements. The differences between values in the third column merely reflect the fact that the data are only accurate to about 15 percent. The values of the numbers themselves are of no consequence, for they depend on the size of the screen and the time spent observing it, which was several hours for the data reported here. But the approximate constancy was significant. It indicated that the enormously varying counting rates were merely multiples of the factor $1/(\sin \frac{1}{2}\theta)^4$. The errors were of no importance. Had Thomson's atom been correct, using the number of flashes at 15° to predict the number of larger angle ones, one would have expected only 16 instead of 1435 at 45°, and one could have waited all day without seeing a flash beyond 75°.

TABLE 14-1

(1) Angle of deflection θ, degrees	(2) Theoretical scattering rate $\dfrac{1}{(\sin \frac{1}{2}\theta)^4}$	(3) Number of flashes observed	(4) Col. 3 ÷ col. 2
150	1.15	33.1	28.8
135	1.38	43.0	31.2
120	1.79	51.9	29.0
105	2.53	69.5	27.5
75	7.25	211	29.1
60	16.0	477	29.8
45	46.6	1,435	30.8
37.5	93.7	3,300	35.3
30	223	7,800	35.0
22.5	690	27,300	39.6
15	3445	132,000	38.4

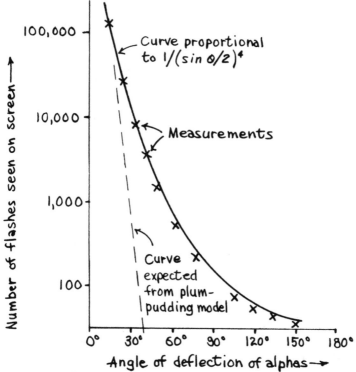

Figure 14-6

The fractional numbers given for some of the high angles reflect the fact that a longer period of observation was required for these measurements, because of the rarity of large-angle scatters; the results are averages of several measurements.

The Geiger-Marsden data serve to illustrate that while physics is often described as a "precise" science, most experiments are no more precise than they need be. The discrepancy between the predictions of the plum-pudding and nuclear atoms was so enormous that a 15 percent measurement was enough to settle the issue. To strive for greater precision would have been a waste of time, especially for one with such a broad range of interests as Rutherford. Further refinements of the experiment were left to less original minds, and no one remembers their names.

The data are graphed in Fig. 14-6 to show the agreement between theory and experiment more vividly. In order to graph such a wide range of values of number of counts, a *logarithmic scale* is used: i.e., the vertical height of each point on the graph represents the logarithm of the number of counts. This has the effect of compressing the larger numbers, so that the region 100 to 1000 occupies no more space on the graph than the region 10 to 100, rather than taking up 10 times as much space. It is a common device of scientists to represent graphically numbers which vary over an enormous range.

Rutherford was able to draw even further conclusions from these data. Coulomb's law and the known velocity of the alphas enabled him to calculate just how close the alphas had come to the nucleus; the alphas that come off in the backward direction have made the closest approaches, as indicated in Fig. 14-5. Had they struck the nucleus, one would expect to find a deviation from the predicted scattering of backward particles, since new forces would have come into play. Rutherford concluded the nucleus must be smaller than 10^{-12} cm, or one ten-thousandth the diameter of the whole atom! As it turns out, the alphas had made a near miss—a gold nucleus is about 10^{-12} cm in diameter.

This is the first time in this book we have presented the full details of an experiment. It was done in part because of the importance of the experiment for all of atomic physics and also because it came in the era when reporting the data and experimental details, rather than merely the conclusions, became standard practice in physics journals. But this particular experiment has further historical significance. It was the first demonstration of the power of the study of particle scattering as a tool for uncovering the nature of forces that operate in the subatomic world. Most of the experiments conducted today with nuclei and subnuclear particles are in a sense variations on a theme by Rutherford, Geiger, and Marsden. Shooting things at atoms and nuclei and studying where they go after the collision is one of the few probes physicists have to get at the workings of matter on the submicroscopic level.

Of course, Rutherford's work raised more questions than it answered. If the positive charge and most of the mass were concentrated in a tiny core of the atom, where were the electrons? Rutherford himself had no idea how to proceed. If the theorists said it couldn't work, let *them* figure out why it worked nonetheless. Rutherford dealt in experimental fact.

But there was at least one young theorist who greeted Rutherford's results with enthusiasm. This enthusiasm would eventually make Niels Bohr the guiding spirit in the development of the new physics, the man who dared to make the final break with three centuries of physical thought.

Bohr had come in the autumn of 1911 to J. J. Thomson's Cavendish Laboratory, to cap his rigorous continental training in theory with an exposure to the British style of experimental physics. Though he arrived highly recommended and bubbling with enthusiasm, this 26-year-old Dane found the Cavendish no bed of roses. First there was the petty arrogance of the Cambridge tradition. His easy-going, boyish character rebelled against the formality of a medieval university town. In a letter home, he complained that his tutor had presented him with "a whole book" on the do's and don't's of academic protocol.

In the laboratory itself Bohr fared little better. The Cavendish had grown in response to Thomson's reputation and was by then far too large for one man to handle. Bohr described the atmosphere of the laboratory as "a state of molecular chaos." Finally, Thomson had shown little interest in Bohr's doctoral thesis, which was based in large measure on Thomson's own work on electrons. Bohr had translated it into English, hoping that his host could help find a publisher that would guarantee a wider audience than was available in Denmark. But it remained on the busy professor's desk, unread.

When Rutherford visited Cambridge for a reunion of Thomson's "old boys," he was introduced to Bohr. By a curious circumstance, the young Dane's name rang a bell. Bohr's brother Harald, who was to become a celebrated mathematician, was at that time better known as the star of the Danish football (soccer) team that had made a surprisingly strong showing in the 1908 olympic games in London. Though Niels had never advanced beyond reserve status on his college team, Rutherford, an avid sports fan, referred to him from that day forward as "that football player." An invitation was extended to visit Rutherford's new Manchester laboratory. Bohr returned to Cambridge with stars in his eyes, preaching the gospel of the planetary atom.

To Thomson this was the last straw. It was bad enough that his most illustrious student was challenging his model of the atom. That Rutherford should seduce a guest at the Cavendish into his heresies was too much. By mutual agreement of all parties, Bohr left in April 1912 to finish out his English sojourn at Manchester.

The industrial Midlands were a far cry from the formality of Cambridge. The laboratory was young, the University was young, and above all Bohr was in day-to-day contact with physicists as young and enthusiastic as himself. This was not a place where one had to apologize for one's crazier ideas; Bohr was in his element.

The unique skill that Bohr brought to Manchester was a thorough knowledge of a hodge-podge of radical new ideas called the *quantum theory*, which was still practically unknown in Britain. Bohr was convinced that in these ideas lay the key that would make the nuclear atom work. To see how right he was, we must now begin the story of the quantum itself. This is the task of the next chapter.

CHAPTER FIFTEEN

The Atom and the Quantum

The universe is not only queerer than we imagine,
but it is queerer than we can imagine.

—J. B. S. HALDANE

ANY PHYSICIST WRITING on the early steps in the development of quantum mechanics is faced with an embarrassment similar to that of a Victorian author writing an uplifting biography of a statesman of illegitimate origins. The problem is to get over the birth and early years as quickly and discreetly as possible, without sowing the seeds of confusion and thereby placing in jeopardy the inspirational message.

For the origins of the quantum theory are obscure and embarrassing indeed. It was bad enough that the idea of discontinuity on the atomic scale should have entered physics through the study of a phenomenon that is neither atomic nor discontinuous. But on top of this, the phenomenon is so difficult and obscure both mathematically and physically that few physicists are at all familiar with it.

PLANCK'S STAB IN THE DARK

The phenomenon that led to the birth of the quantum theory is that of the continuous spectrum of light emitted by densely packed matter heated to incandescence, as in the filament of a light bulb or a bar of red-hot steel. The phenomenon is not atomic, because the atoms are closely packed and in continuous interaction with one another, completely disrupting their natural way of producing light. And it is certainly continuous: light is emitted at all wavelengths, the relative brightness of the different colors depending solely on the temperature of the material. Extensive experimental studies were conducted in the last two decades of the nineteenth century, finding the brightness (energy radiated) at various frequencies. A typical result is shown in Fig. 15-1.

Understanding this spectrum was viewed as a problem in thermodynamics, the theory of the conversion of heat into other forms of energy, which was then perhaps the most sophisticated and highly developed branch of physics. The theory had successfully explained the conversion of heat into mechanical energy in engines and chemical energy in chemical reactions, and a host of other phenomena. And it had shown that the detailed properties of the materials involved in these transformations were of little importance, which also seemed to be the case with incandescent-light emission. A bar of any red-hot metal looks much the same as any other. Surely thermodynamics was the right theory to cope with the conversion of heat into light.

The first attempts showed encouraging results. It proved simple to account for the fact that the total energy radiated increased as the fourth power of the absolute temperature; the theory also satisfactorily

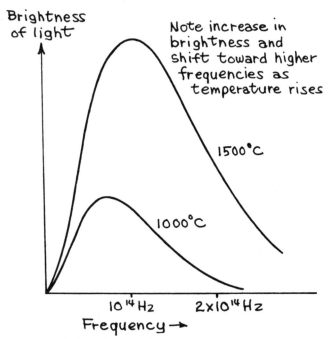

Figure 15-1 Incandescent-light spectrum.

explained why when an object is heated, the color of light emitted changes from a dull red through orange to white and on to blue, as the temperature rises. But the exact spectrum shown in Fig. 15-1 eluded mathematical description.

Just before the turn of the century, Max Planck, a young German theorist, undertook the study of the problem. With the common sense that was typical of his approach, Planck realized that most mathematically difficult problems are solved only when the correct answer has already been guessed. Accordingly, he set out, by a process of trial and error guided by the successes and failures of previous attempts at the problem, to find a formula that would fit the experimental curve in Fig. 15-1.

His search was rewarded early in the first year of the new century. But, as Planck fully realized, finding an empirical formula, one that fits the data but is not based on a physical picture of the process being described, can never be an end in itself in physics. Without an explanation of its underlying significance, the formula would remain an obscure curiosity.

By the end of the year, Planck had found a supporting argument for his lucky guess, but it involved a peculiar assumption. He found he was forced to assume that the conversion of heat into light could not occur in any amount whatsoever, but came in the form of chunks whose size depended on the frequency of the light produced. The smallest amount of heat energy that could be converted to light of frequency v was given by the formula

$$E = hv \qquad (15\text{-}1)$$

and only whole-number multiples of this amount could be produced. The constant h appeared in Planck's empirical formula. It is known to this day as *Planck's constant* and ranks with the velocity of light as one of the two basic constants of nature, though its significance is perhaps somewhat more mysterious than that of c. The introduction of this constant and Eq. (15-1) before a meeting of the German Physical Society on December 14, 1900, is usually taken as the birthdate of the quantum theory.

Nothing could have been more unnatural in the physical thought of the time than to arbitrarily restrict energy changes to granular *quanta,* as Planck called them. Not even the author of the assumption was sure it should be taken seriously. Perhaps there was some way to explain or eliminate this granularity. For example, the empirical formula might not be correct; after all, experimental data are never perfect. A slightly different formula, derived without the assumption, might fit the data just as well. Or perhaps a different route to the same formula could be found, one that did not involve such an arbitrary and unreasonable assumption. The theory of heat was so mathematically involved that no one could be sure. At the worst, if the assumption held up, it might represent some obscure property of the electron, which after all had to do the job of energy transformation, since light can only be produced by a moving electric charge and the lightweight mobile electron was the ideal candidate. The idea of the quantum was regarded as no more than a way station on the route to a deeper understanding, and physicists working on the problem assumed it would eventually disappear.

But both the formula and the idea of the quantum survive to this day. Indeed, it is the fact that the cosmic "background" signal of millimeter-length waves exactly follows the Planck formula for a body of temperature 3.2K that provides the principal experimental prop for the "big bang" theory of cosmology.

EINSTEIN AND 1905 AGAIN

There are always a few bold thinkers who, confronted with a puzzling new idea, will look not for ways around it but for new ways to extend and test it, to see if it crops up somewhere else in nature. Hardly anyone was bolder than Einstein in 1905, the year in which he announced the theory of relativity. The problem Einstein dealt with was not, like Planck's, a heavily studied phenomenon right at center stage in the attention of physicists, but a relatively obscure one called the *photoelectric effect.*

A series of accidental discoveries late in the nineteenth century had led to the conclusion that light was capable of knocking electrons out of metals. This surprised no one, for light being an oscillatory electromagnetic field, it was bound to give an electron a good shaking and might actually shake it loose from its moorings. Few experiments had been undertaken on this relatively obscure effect, but some qualitative facts were known about it, and these had no obvious explanation:

1. The effect was easy to produce with blue or ultraviolet light but not with red light.

2. The rate at which electrons were given off was proportional to the brightness of the light.

3. If one measured the energies of the electrons given off, however, they proved to be independent of the brightness of the light, but there were indications that the energy did depend on the wavelength.

The apparatus for observing the photoelectric effect is depicted in Fig. 15-2. Light shining on the photocathode liberates electrons, which are collected at the anode, producing a measurable electric current. By

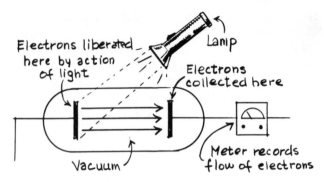

Electrons liberated here by action of light

Lamp

Electrons collected here

Vacuum

Meter records flow of electrons

Figure 15-2 Photoelectric effect.

charging the anode negatively, so that it repels electrons, the current can be stopped. From the size of the force required to repel the electrons, their energies can be calculated.

The third fact above was the key clue and was in part another contribution of Ernest Rutherford, from his last paper as a research student at Cambridge before departing for Montreal. It was certainly the most peculiar fact of all. By simple reasoning, brighter light meant stronger electric fields, which in turn meant greater forces on the electrons. Surely they must come out faster. Yet the experiments were still crude, and it seemed premature to get unduly alarmed about a few peculiarities.

Einstein had been worrying about what seemed to him an awkward point in Planck's theory. How he even found time to think about it, in view of his other preoccupations, is somewhat of a mystery. Planck had assumed that the emission of light in quanta was due to limitations on the motion of the oscillating electrons that produced the light; yet otherwise he used the laws of classical physics to deal with the process. Since there were no other examples of limits on motion outside the requirements of Newton's laws, to Einstein it seemed more logically consistent to ascribe this peculiar behavior to the light itself. Once he came to this point of view, he began to return to a particle picture of light to explain Planck's quanta. Searching the experimental literature for some way to test his conjecture, he found that the few known experimental facts about the photoelectric effect seemed to lend themselves naturally to a particle theory, in which light would be a hail of quanta, today called *photons*. The most puzzling feature, point 3 above, was then naturally explainable. An electron could at most absorb a single photon—it would be removed from the metal so fast that in light beams of normal strength it would never get a chance to trap another. Thus, the energy of the electron would depend solely on the energy of the photon, and therefore on the frequency (color) of the light. It would not depend on the brightness, which was merely a measure of how many photons were available. More photons meant more electrons, but not faster ones, which explained point 2 on page 183.

Point 1 also seemed reasonable in a quantum picture. A certain amount of energy is required to pull an electron free of the forces of attraction that hold it in the metal. If one photon didn't provide enough energy, the electron was stuck. It would quickly lose its added energy before another photon came by. And red light, being lower in frequency than blue, had less energetic photons.

Like any good theory, Einstein's picture of the photoelectric effect permitted a conclusive experimental test. The maximum possible energy of the electrons should be $h\nu$ minus the energy lost in pulling an electron loose. But h was a constant obtained from an empirical formula for incandescent light! If it turned up again in the photoelectric effect, the coincidence would be too much to swallow; Einstein must be right. If the same granularity appeared when light was created and when it was absorbed, it seemed more natural to ascribe the granularity to the light itself than to the two unrelated processes that produced and absorbed it. But the experiment was a difficult one, and 11 years were to pass before the American physicist Robert Millikan confirmed the prediction.

The prediction confirmed by Millikan is depcited graphically in Fig. 15-3. Until the frequency gets high enough for a single quantum to free an electron from the forces that bind it to the metal, no electrons are liberated. From then on, the electron energy is equal to the quantum energy diminished by the amount lost in overcoming the binding force. The crucial quantitative point is that the slope of the graph, which is the ratio of the increase in electron energy to the increase in frequency, is Planck's constant h, even if the process that produced the light has

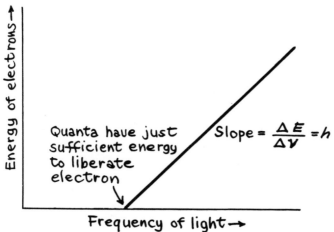

Figure 15-3 Einstein's prediction of the relationship between electron energy and light frequency in the photoelectric effect.

nothing to do with the process described by Planck. Thus, the relationship $E = h\nu$ must pertain to light itself.

Until the experimental confirmation was forthcoming, the theory of the photoelectric effect remained the least respectable of Einstein's many achievements. How could it be reconciled with the overwhelming evidence for the wave theory of light? His quanta of light must obviously be more like particles than waves, so how could they produce all the wave effects that had been observed in the century since the work of Young? How could light be *both* a particle and a wave at the same time? Einstein's attempts to struggle with this dual character of light, today a central concept in the quantum theory, reassured nobody. When he was put up for membership in the Prussian Academy of Sciences in 1913, Einstein's sponsors felt compelled to excuse this peculiar lapse of an otherwise obviously brilliant man. But only 3 years later Millikan's work was to show that Einstein was devastatingly right in this, his least solid contribution. The photoelectric effect, rather than the theory of relativity, was the basis for awarding Einstein the 1921 Nobel prize in physics, for relativity was still a bit controversial at that time.

But the spirit of the time was receptive to new ideas. With young people seeking new directions in the world of art, music, and literature, Einstein and his café companions had learned to question established ideas in politics, morals, the arts. Why not try it in physics? To a new generation of physicists, Einstein's photoelectric theory was a hope rather than a problem; clearly, there was more to Planck's quantum than some obscure feature of the theory of heat.

Their most urgent need was for a natural way to slip the quantum into the laws of motion. Fortunately, h itself obligingly suggests a way to do this. The hint is in the physical units in which it is expressed. They are the units of a quantity called *action,* which was coming to be regarded as possibly more important than energy or momentum themselves. Action is the product of *momentum and distance moved,* or alternatively *energy and time of travel.* Though not itself necessarily a conserved quantity, it does provide an umbrella that covers both conservation laws.

In the mid-nineteenth century William Rowan Hamilton, Astronomer Royal of Ireland, had shown that action could be used to simplify difficult problems in orbit theory. He proved that once two points on an object's path are known, the route between is always *whichever one makes the action smallest.* This provided the "young Turks" of the quantum revolution with a pocket tailor-made for Planck's constant.

BOHR ASKS THE RIGHT QUESTION

It was a representative of the new breed of theorists that Niels Bohr arrived at Rutherford's Manchester laboratory in the spring of 1912. He had been raised in the comfortable, supportive environment of one of Denmark's leading intellectual families. The Bohr home had been the social center for a lively circle of philosophers, scientists, and writers, and the precocious Bohr children were encouraged to participate in their discussions. Coming as they do from a small nation with a difficult language, educated Danes usually try to keep one foot in the English-speaking and the other in the German-speaking world, so young Niels was exposed to as broad a range of ideas as could be found anywhere in the world.

The Danes also preserve the charming custom of holding oral examinations for the doctorate as public events. Niels and Harald defended their theses before packed houses, to rave reviews in the daily press! With such a background, Niels Bohr never hesitated to believe he could turn the world upside down with the products of his mind.

Rutherford was bound to be flattered that such a well-recommended young man viewed his nuclear atom with enthusiasm, rather than the dismay expressed by most of his own contemporaries. But an intuitive thinker like Rutherford, who liked to visualize the unseen objects he worked with, was bound to view Bohr's interest in quanta with some skepticism. Rutherford had once answered a colleague's dinner-table question whether he actually thought electrons and alpha particles really existed with the reproof: "Not exist—not exist—why, I can see the little beggars there in front of me as plainly as I can see that spoon!" Quanta were an entirely different matter, a bit too abstract for Rutherford's taste. Yet he took a personal liking to the eager young Dane, explaining this deviation from a proper British disdain for theorists (and especially continentals) with the remark, "Bohr's different. He's a football player!"

Others before had applied Einstein's and Planck's quantum concept to the atom, usually in the context of the more popular Thomson model. Most of these studies had been efforts to explain away Planck's constant by showing it to be a consequence of the size of the atom and the charge and mass of the electron. Bohr immediately saw that in dealing with a nuclear atom, the correct approach must be precisely in the opposite direction. Since in the classical theory electron orbits could be any size whatsoever, some new principle must be found to explain atomic size, and perhaps this principle could get rid of the "death spiral." It was not the duty of the atom to explain Planck's constant but for the constant to explain the size and stability of the atom.

This insight was a fortuitous bonus of the research institute structure mentioned in the preceding chapter, in which the best young physicists traveled a great deal. Bohr's continental education had given him the required familiarity with the quantum theory; but only at Manchester could he find people who really believed in the nuclear atom and would encourage his efforts. Despite the existence of scientific

Bohr's engagement photo, taken shortly before his departure for England. (Courtesy of American Institute of Physics.)

journals, the emotional impact of scientific ideas is always greatest at their source. Elsewhere in the world, Rutherford's nuclear atom seemed a possibility. At Manchester, Bohr was in the midst of a group of talented young physicists who took it as established fact. While at the Cavendish or in Copenhagen Bohr might have toyed with the idea that the quantum theory might permit a stable planetary atom. At Manchester, day-to-day results from the darkened laboratory down the hall and the enthusiasm with which his coworkers received them changed the whole picture. The planetary atom *had* to be made to work, and all the signs were that the quantum theory held the key.

The work of Planck and Einstein inseparably tied the frequency of light to its energy. Bohr reasoned that this implied that an atom can only exist in a limited set of *states* of internal motion, each with a definite energy. Light is emitted when an atom changes it state, with the frequency determined by the difference in energy between the initial and final states.

In a planetary atom, the energy of internal motion depends solely on the size of the orbit. Bohr realized he must find a new quantum rule that restricted the motion of electrons to a limited set of orbits. The Einstein-Planck rule $E = h\nu$ was not enough; if the electrons were free to move in any way they chose, all emergies would be possible, and thus all frequencies of light would be emitted. Only by restricting the orbits could a line spectrum be obtained.

This restriction could also account for the stability of atoms; once in the smallest orbit, the electron would have no place to go. A collision between atoms could knock it into a higher energy orbit, and it would emit light once or several times in the process of returning to the smallest orbit, or *ground state,* in which it normally existed.

Bohr was forced to assume that the electrons in the legal orbits would ignore the classical laws of electromagnetism and radiate no light. Otherwise there would be a gradual loss of energy, inconsistent with both the quantum picture of light and the rule restricting orbits. Only while shifting orbits could the electron produce light.

He also guessed that the orbit rule would somehow involve Planck's constant. Some sort of constant was needed to set the scale of the orbits, and he hoped he could use Planck's and not have to introduce a new one.

Once the problem had been thus formulated, the task was clear: find the orbit rule. But his scholarship to England had run out. He was married upon his return to Copenhagen, and a honeymoon in Norway took care of the rest of the summer of 1912. Starting his first teaching job in the fall, he had little time to concentrate on the problem. He played with orbit calculations, but with no experimental data to guide him, he bogged down. It was only his ignorance of atomic spectra that hampered him. He had asked the right question; he was soon to find that nature had already given the answer.

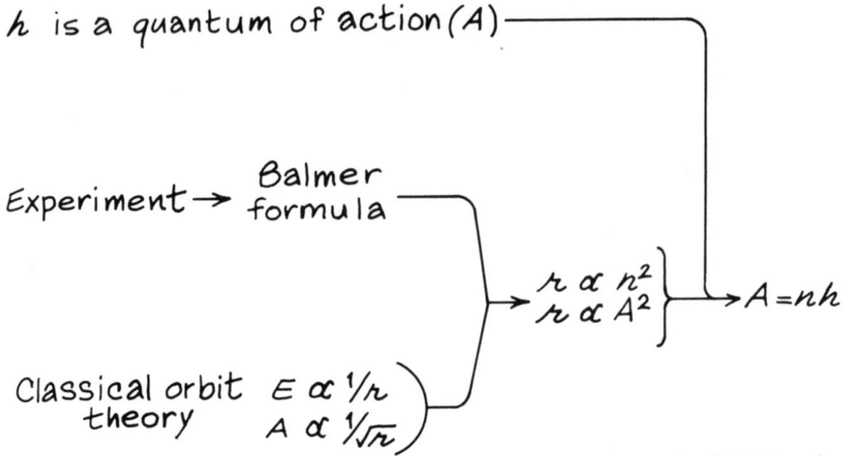

h is a quantum of action (A) ———————

Experiment → Balmer formula ———

Classical orbit theory $E \propto 1/r$ $A \propto 1/\sqrt{r}$

$r \propto n^2$
$r \propto A^2$ } → $A = nh$

Figure 15-4 Logic of Bohr's rule.

WHERE THE ACTION IS

At the end of January 1913, H. M. Hansen, an old classmate of Bohr's and a spectroscopist, paid a visit to Copenhagen. When he asked Bohr if the theory on which he was working might explain the Balmer formula, the reply was, "What formula?" He advised Bohr to look it up. The moment Bohr set eyes on the formula, he knew his search was over. The allowed orbits were circles for which *the action is an integer multiple of h.*

To anyone familiar as Bohr was with orbit theory, the reasoning that led to this rule was simple. The chain of logic is outlined in Fig. 15-4.

The Balmer formula gave the lines in the hydrogen spectrum in terms of differences of inverse squares of whole numbers. Bohr knew the frequencies of spectrum lines must depend on differences in emergies of orbits, so the formula immediately suggested that the allowed orbits had energies that followed the same proportionality; i.e., they went in the series $1, \frac{1}{4}, \frac{1}{9}, \frac{1}{16}, \ldots$. From playing around with circular orbits, Bohr knew well of two mathematical properties they obeyed: (1) the energy is inversely proportional to the radius, and (2) the total action is proportional to the square root of the radius. The first property, taken with Balmer's formula, indicated that the radii of successive orbits were 4, 9, 16, ... times the radius of the inner orbit, as shown in Fig. 15-5.* Then the second told him the action of these orbits went as the whole numbers 1, 2, 3, 4, ...! The last piece in the jigsaw puzzle was in place, for the unit by which the action increased turned out to be *h!*

---

* If it appears that the energy is *highest* for the small orbits, remember that all these energies are *negative*. That is, the (negative) potential energy exceeds the kinetic, and the orbits are bound, as explained in Chap. 6.

Of course, the rule was completely ad hoc. Nothing in the work of Planck or Einstein even hinted that Planck's h could be used for such an outrageous purpose. And Bohr couldn't pretend to describe how an electron jumped from orbit to orbit or why it didn't radiate light when it stayed in an orbit. As he himself explained it, the orbit rule was probably not the *real* rule. It was just an expression *in the language of the old physics* for a rule whose meaning would become clear when the true language of quantum mechanics was discovered. The next chapter will show just how right this impression was.

Bohr fully realized he was treading on thin ice. The old physics had three centuries of work to back it up; all he had was Rutherford's experiments and the Balmer formula. Somehow he had to work out a way for his fledgling theory to coexist with the old physics. This he provided for by means of his *correspondence principle*. The quantum theory would reign supreme on the atomic scale, classical physics on the gross scale of daily experience. Where they overlap, both theories must give the same answers.

For example, consider some of the very largest orbits of the hydrogen atom. Orbit no. 2 is four times as large as no. 1. But orbits no. 10,000,000,000 and 10,000,000,001 would be very close in size. The jumps between such outer orbits would be almost unnoticeable; the process might well closely resemble the continuous spiraling in the classical theory. Bohr showed that for two adjoining large orbits, the

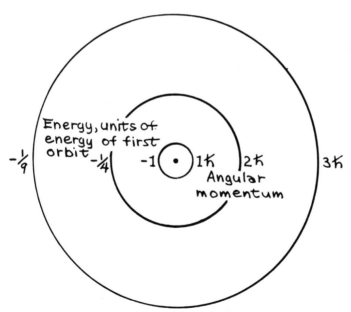

Figure 15-5 First three Bohr orbits.

light emitted in a quantum jump has the *same frequency as the orbital revolution*, which is what classical physics would predict. He used this slim reed to lend credibility to the theory.

In many formulas derived from Bohr's orbit rule, h turns up divided by 2π. To simplify these formulas, $h/2\pi$ is given a symbol of its own $\hbar$, pronounced "h bar."

NUMBERS ARE POWERFUL CONVINCERS

Outrageous as it might be, nobody could dispute the fact that Bohr's theory came up with those marvelous little convincers, experimental numbers. Frequencies of spectrum lines can be measured to fantastic accuracy, and Bohr hit them right on the head. Let us summarize his quantitative results. To avoid interrupting the flow of the narrative, we relegate the simple algebra on which they are based to the section of the Appendix for this chapter.

The radius of the first orbit is given by the expression

$$r = \frac{\hbar^2}{me^2} = 0.53 \times 10^{-8}\,\text{cm}$$

where m and e are the mass and electric charge of the electron. The energy of this orbit is

$$E_1 = \frac{me^4}{2\hbar^2}$$

The point of displaying these curious expressions is that they are, in fact, peculiar looking combinations of tiny quantities. That they turn out to give the right answer could hardly be a coincidence. The radius jibes with the known size of the hydrogen atom; the energy, divided by h, gives the constant in Balmer's formula, which was known to high precision.

The theory also explained the Ritz principle in a satisfying fashion, as illustrated in Fig. 15-6. If the frequency of a spectral line turns out to be the sum of two others, it merely means that the line comes from a big jump that skips one or more orbits; the other two lines complete the same transition in two steps. Bohr's work suggested that the Ritz rule might apply to sums of three or more frequencies, and examples were quickly found.

The Bohr theory also explained one curious fact about atomic spectra: atoms cannot absorb light at all the frequencies they emit. In the Thomson atom, light whose frequency matched the natural frequency of oscillation of one of the electrons would set the electron in motion,

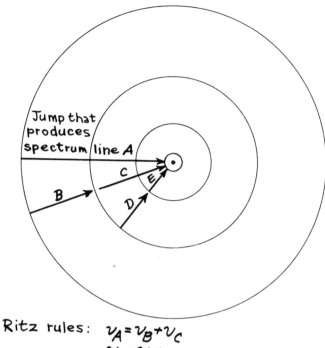

Ritz rules: $\nu_A = \nu_B + \nu_C$
$\nu_C = \nu_D + \nu_E$

Figure 15-6 Bohr's explanation of the Ritz principle.

soaking up the energy of the light, a natural reversal of the process of light emission. In the Bohr theory, since undisturbed electrons are always in the lowest orbit, only light of the frequencies produced by jumps from higher orbits to the lowest could be absorbed. In Bohr's scheme, light absorption is also the reverse of light emission; a photon is "sucked in," and the electron jumps to a higher orbit. But if any photon came by whose frequency corresponded to a jump between two higher orbits, it would be ignored, because in undisturbed atoms these orbits are empty.

A more direct confirmation of Bohr's outrageous ideas was soon to come. The experimenters Franck and Hertz, in 1914, studied collisions between free electrons and gas atoms. They found that the energy losses suffered by the electrons exactly matched the energies of light quanta produced in the spectrum of the gas by the formula $E = h\nu$. If electrons were used that had too little energy to cause the first quantum jump, all collisions were elastic: no energy was lost. Once they were raised above this energy, in some of their collisions they would lose exactly enough energy to produce one quantum of light. What was happening, of course, was that the kinetic energy of the free-flying electron was being used to raise an orbiting electron in the atom to a higher orbit. This was finally a direct *mechanical* confirmation that the basic Bohr idea was right. The internal motion of electrons in atoms only permitted a limited set of energies. Even when light was not involved at all, Planck's constant came into play.

Like the Ritz principle, the Franck-Hertz result held for gases other than hydrogen, though there was no formula for the energies of these orbits—they had to be inferred from the observed line spectra. Thus, there was reason to believe Bohr's ideas could be extended to more complex atoms, with many electrons.

THE BATTLE IS JOINED

Bohr's theory was greeted with considerable skepticism. Some sort of fixed-energy-level picture of the atom was obviously called for, but was Bohr's absurd picture the right one? As Bohr himself realized, his theory was only a step on the route to the truth. Rutherford, for all his dedication to the nuclear atom, was afraid to give it much credence. Bohr was onto something big, but clearly more work had to be done.

And suddenly it was August 1914, and there was nobody left to do the work. With the assassination of an Austrian archduke in the obscure Bosnian city of Sarajevo, young scientists who had been working side by side suddenly found themselves conscripted into rival armies, to face one another across fields of barbed wire.

Rutherford's Manchester laboratory was quickly denuded of his "boys," the young men who had made it hum. Rutherford patriotically resigned himself to research on submarine detection, while his friends and former coworkers in the German laboratories worked to frustrate his efforts. Rutherford had good personal reasons to reflect bitterly on the futility of the slaughter. His own brilliant student and protégé, Henry Moseley, died in the disastrous British attack on the Dardanelles. Moseley's greatest achievement, just before he left for the army, had been to show by experiment that the Bohr theory explained the production of x-rays. But this did not exempt him from the horror that cut down a major share of European youth of his age.

PHYSICS IN THE TWENTIES

When the surviving young physicists returned from the trenches at the end of 1918, the intellectual world was in no mood for caution. The old political order had fallen in the defeated Central Powers and had been discredited in the eyes of many intellectuals among the victorious allies. The Bolshevik revolution in Russia had aroused the passions of the whole world.

Physics often seems the most insulated of intellectual endeavors. Few historians of science ever concern themselves with anything but its internal history. Perhaps it was a mere coincidence that classical physics seemed on the verge of ultimate triumph, in the complacent 1880s, only to be replaced by a new physics born in the turbulent 1920s; in purely scientific terms, the time was ripe. Not all sciences experienced a similar flowering. And great physics is usually done by people who, for the time being at least, are thinking of little else.

Yet one can imagine the mood of a young physicist whose diversions might include the plays of Brecht, the music of Hindemith, the novels of Thomas Mann, and Dada art exhibits. His neighbors might be surrealist painters, radical poets, Bauhaus architects, or devotees of the daring psychological theories of Jung. The mood of the cafés quickened the pulse. Was this the time to dwell contentedly on the ancient traditions of physics, to add one's little bit to the great edifice built on the foundations of Newton? The young Fellow or *Privatdozent* could hardly face his friends if he engaged in a pursuit that seemed so sterile by their values.

Science had ridden high in imperial Germany, a state that claimed to rest its legitimacy on scientific principle. In the new Weimar Republic, its status was suspect. Classical physics was in particular disfavor due to the role assigned it in one of the most influential books of the era, Oswald Spengler's *Decline of the West.*

Spengler condemned Western civilization for the faustian bargain it had made for mastery over nature. The dead hand of determinism, epitomized by the laws of classical physics, had traced the West's signature in blood. Echoing the romanticism of a century earlier, Spengler proclaimed *mysticism, intuition,* and *chance* the stuff of life on which a new civilization would arise.

The quantum chasers did not take this criticism lying down. There was mystery aplenty in quantum jumps and intuition in their methods. A new physics was aborning, and determinism was finished.

Einstein and Planck were wary of the new mood. By all means build the new physics, they warned, but to capitulate to the forces of unreason was to invite in the hounds of hell. Already these were very much in evidence on the streets, with their brown shirts and swastika armbands, and listed among their targets was the "Jewish science" of theoretical physics.

But in this inhospitable intellectual climate, German physics rose to its greatest heights. In the eye of the hurricane sat the great Georgia Augusta University at Göttingen, a provincial university city reminiscent of American college towns. Göttingen had been a center of intellectual ferment and political rebellion since its foundation by George II, Elector of Hanover and later King of Great Britain. The cogent fact in the twenties was the presence of David Hilbert, perhaps the most influential mathematician of our century. He created an atmosphere of intense scholarly debate that cut across traditional disciplinary boundaries. Hilbert's efforts had been aided by historical accident—a prize left to the university for a proof of one of the famous unsolved problems of mathematics, Fermat's theorem.* With no claimants for the prize, Hilbert used the income from the legacy to import distinguished lecturers to the otherwise isolated Hanoverian town. Hilbert chose his speakers judiciously, to maximize the atmosphere of controversy. The roster includes quite a list of Nobel laureates—who were usually invited *before* the award of the coveted prize. A speaker approaching a Göttingen lecture knew he faced an audience that would accept nothing less than his most original ideas and that he must be prepared to defend his work against sharp attack. There were debates, long into the night, on the shape of the new physics. The students who flocked to Göttingen were in no mood to be polite to their elders.

But the mecca of this new religion was unquestionably Copenhagen, and its unchallenged prophet was Niels Bohr. Barely into his thirties, Bohr headed a new institute supported in part by the profits of the venerable and world-renowned Carlsberg brewery. Bohr was to make Copenhagen nearly as famous for physics as for beer and pretty girls. Rumors of an important new development anywhere in Europe brought a reflex reaction to the young guard of physics—catch the next train to the charming and fun-loving city on the Öresund. Only in "The Presence" could the true significance of a new idea be evaluated, and the debates became legendary.

Bohr's work habits enhanced his influence. Unlike Einstein, who craved solitude, Bohr thought best out loud, in a madcap environment full of people to bounce ideas off of. Work at his Institute might be conducted across a ping-pong table (Bohr was almost unbeatable) or on a tour through the Tivoli amusement park. Bohr's ideas often began as metaphors, poetic images with meanings his listeners could scarcely fathom, but which encouraged them to let their own imaginations run free. In Copenhagen, the new physics was not so much *built* as slapped together in a riotous spree of individual and collective creative effort.

The settings were familiar—the Institutes of the German and French universities, the Cavendish Laboratory, a bit less stuffy after Rutherford succeeded J. J. Thomson in 1919, the gentle Scandinavian

---

* The mathematician Pierre de Fermat (1601–1665) stated in a marginal note found after his death that the equation

$$x^n + y^n = z^n$$

has no whole-number solutions for n greater than 2 and that the proof of this was "obvious." It couldn't have been *terribly* obvious, for mathematicians have been looking for the proof to no avail in the three centuries since.

frivolity of Copenhagen. But the mood was new, and every bit as romantic as Hemingway's Paris or Brecht's Berlin. In Munich, in the same era that witnessed Hitler's beer hall *Putsch,* the waiters at one café near theorist Arnold Sommerfeld's *Institut* had peculiar instructions. When the young physicists who passed their evenings there left the marble-topped tables covered with equations, they were under no circumstances to be cleaned; a number of the key ideas of the new physics went straight from those tables to the pages of *the* journal of the quantum-mechanical revolution, *Zeitschrift für Physik.*

But the spirit of the era is epitomized in the unforgettable image of Fritz Houtermans, who first untangled the chain of nuclear reactions that provides the sun with its energy. Few scientists have ever succeeded in fully articulating the terror and beauty of the feeling that comes with the realization that one has solved an age-old mystery. In the height of this mood, the day the last piece of the puzzle fell into place, Houtermans had a date. As they walked the quiet, dark streets of Göttingen, his fraulein, sensing the exaltation of her escort's mood, remarked on the beauty of the starry sky. "Yes," the young theorist replied, "and would you have guessed that you were arm in arm with the only man alive who knows why they shine?"

CHAPTER SIXTEEN

When is a Particle a Wave?

Hail to Niels Bohr from the worshipful Nations!
You are the Master by whom we are led,
Awed by your cryptic and proud affirmations,
Each of us, driven half out of his head,
    Yet remains true to you,
    Wouldn't say boo to you,
Swallows your theories from alpha to zed,
    Even if—(Drink to him,
    Tankards must clink to him!)
None of us fathoms a word you have said!

—ANONYMOUS (POSSIBLY GEORGE GAMOW)

POPULAR ACCOUNTS OF scientific discoveries often describe a research scientist as a hunter stalking his prey. The young men who built quantum mechanics in the early 1920s more nearly resembled a rowdy band of schoolboys chasing a rabbit across a rocky meadow. The order of the day was "anything goes," and the approach was scandalously empirical. Invent a quantum rule, as Bohr did, derive a formula, fit the data, and go on to the next problem. An understanding of what sort of physical reality might underlie a successful computation would come in its own good time. The hope was that after enough lucky guesses a pattern might emerge to guide physicists to a deeper level. In the meantime, the new game was just plain fun.

THE BOHR ATOM IS NOT ENOUGH

The starting point for postwar quantum physics was tinkering with Bohr's model of the atom to make it work for heavier elements. This was no mean task; no longer could one work with a simple attractive force directed toward the nucleus. The mutual repulsion of the electrons had to be taken into account. How this might modify the orbits or affect the quantum rule that led to them was impossible to say on the basis of Bohr's ad hoc and patently incomplete rules. The scheme became more complex. A new quantum rule was concocted by Arnold Sommerfeld that gave each circular orbit a few elliptical companions, each as long as the diameter of the circle. Such an orbit had the same energy as the circular one. But the scheme remained as arbitrary as and more complex than Bohr's original model. Then it became necessary to arbitrarily limit the population of the orbits: two electrons could fit in an orbit, and no more. Where would all this arbitrary rule-making end? It seemed that every new problem brought into being a new rule. It was great fun, but was it physics?

Still, the scheme had some appealing features. Most notable was the fact that it seemed to explain, at least in a loose qualitative fashion, the chemists' periodic table of the elements. Every atom's "ambassadors" to the other world would be its outermost electrons. Its ability to form chemical compounds would depend on their behavior. The theory seemed to suggest that elements in the same column of the periodic table were similar in their outer orbital patterns, and thus the atoms should be chemically similar. There must be some truth in a picture that could shed light on this old mystery. What lay behind all these strange, arbitrary rules?

As is often the case when a group of scientists are caught up in trial-and-error speculation, the key to the muddle came from outside, from a more isolated thinker with the leisure to contemplate the problem in a detached manner. And it came not from Germany, but from France, where a less frenzied and more abstract type of speculation had long been the dominant style in physics. The protagonist, Prince Louis Victor de Broglie, was one of the most improbable characters in a drama full of improbable people.

A PRINCE HAS A CRAZY IDEA

Few families in Europe outrank the de Broglies in the *Almanach de Gotha*, the quasi-official register of nobility. The eminence of his lineage is attested to by the fact that Louis bore the august title of *Prince* as a mere cadet honor; today, following the death of his older brother, he has assumed the higher title of *Duc*. The de Broglies have provided France with diplomats, cabinet ministers, and generals for centuries. One of Louis de Broglie's ancestors even fought on the American side in our War of Independence. Accordingly, de Broglie received the standard humanist education that is the traditional preparation for a role in the French ruling elite.

De Broglie. (Courtesy of American Institute of Physics.)

But the family also had a modest scientific tradition. Louis' elder brother Maurice was an experimental physicist of no mean reputation. Through his brother's influence, the young prince, newly matriculated in 1910 at the University of Paris, took an interest in the work of Einstein. He was particularly intrigued by the possibility of finding a connection between Einstein's two most celebrated works. He speculated that relativity itself might shed some light on the problem of the dual wave-particle character of light in Einstein's work on the photoelectric effect. Work in this field naturally went slowly for this rank amateur, who had in effect to start his education all over again to get into it. The war interrupted his studies before he had made much progress.

Returning from his military service determined to see his interest in physics through to the doctorate, de Broglie retained his preoccupation with the wave-particle problem. But in the meantime, the success of the Bohr model had changed the whole picture in the quantum theory. No longer could quantum effects be regarded as a mere peculiarity of light. The strange limits on the motion of electrons in atomic orbits were an even more disturbing mystery. Convinced that the wave-particle duality was the key to the earlier quantum theory, de Broglie wondered whether it was also the source of Bohr's rules. If light could be a particle, why could not an electron be a wave?

The idea had many inviting aspects. While it was difficult to imagine a mechanical law that would rule out all but a few orbits, similar restrictions on wave motion were quite normal. After all, musical instruments work precisely because the types of wave motion possible on a taut string or in an enclosed column of air are severely restricted. If the electron could in some sense be regarded as a wave, perhaps the Bohr orbits would prove to be standing waves and the atom not so much restricted as *tuned.*

But describing an electron as a wave was quite a bit more difficult than describing light as a particle. Light always travels at the same velocity; thus, its frequency and wavelength are closely related. Once one is known, the other is known from the relation given in Eq. (7-1). For an electron, capable of moving at any velocity whatsoever, there would of necessity be separate rules for the wavelength and the frequency.

Again, it was to his knowledge of relativity that de Broglie turned. In his prewar speculations on the dual wave-particle character of light, he had succeeded in proving that the only way light could be a particle and still move always at the same velocity was for the photon to have zero *rest mass.* In this way, *all* the mass would rise from its motion, and the motion would always be at the velocity of light.

If you find the concept of a particle with zero rest mass disturbing, imagine the photon to be a particle with a very tiny rest mass. Then giving it even a very small amount of energy, as long as it was much greater than the rest-mass energy of the photon, would bring it up to a speed very close to that of light. Most of its mass would be from the added energy; the small rest mass would be unnoticeable, and the small difference in speed of light of different energies (frequencies) would be impossible to measure.

Since the speed of radio waves, which are much lower in frequency than visible light, proves to be the same as that of light, we know that the photon could at most have a tiny rest mass indeed: about 10^{-20} as large as that of the electron.

But a material particle does *not* travel at the speed of light. Knowing the frequency of the electron wave thus did not automatically tell you its wavelength. And to test his surmise that Bohr orbits were standing waves, de Broglie must know the wavelength of the wave associated with the electron. A new rule had to be obtained; $E = hv$ may be enough for the photon, but a separate rule giving the wavelength was essential if the electron were to be described as a wave.

Here his familiarity with relativity proved de Broglie's foremost asset. He noticed that the *wave number* bears the same relation to the *frequency* of a wave that the momentum bears to the energy in the relativistic description of a particle. Both obey triangle relationships of space and time, as illustrated in Fig. 16-1. Here was the key to the problem: if the frequency of a wave is related to the energy of a particle, as in Einstein's description of the photon, perhaps the wave number is related to the momentum. Of course, this could only be a surmise; the argument was an analogy rather than a rigorous derivation.

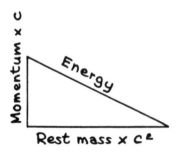

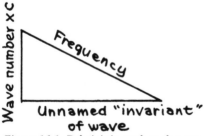

Figure 16-1 Relativistic analogy that suggested de Broglie's hypothesis.

Accordingly, de Broglie set forth, in his celebrated 1924 doctoral thesis, the following formula:

$$p = hk \qquad (16\text{-}1)$$

relating the momentum p of an electron to the wave number k of a wave associated in some mysterious way with the electron.

Now de Broglie could test his surmise about Bohr's orbits. If they were really standing electron waves, the Bohr orbits must be the ones in which a whole number of wavelengths fit, as shown in Fig. 16-2. In this way, the electron wave could travel around the orbit reinforcing itself constructively at each turn, just as the wave on a music string is constructively reinforced by its successive reflections. The waves must fit evenly into the circumference of the orbits. Stated mathematically,

$$\frac{\text{Circumference}}{\text{Wavelength}} = \frac{2\pi r}{\lambda} = n \qquad (16\text{-}2)$$

where n is a whole number. If de Broglie's surmise were to work, this formula must be shown to be equivalent to the Bohr orbit rule, when wavelength is replaced by momentum in accord with Eq. (16-1). Since Eq. (16-1) tells us that $\lambda = h/p$, we can substitute in Eq. (16-2),

$$\frac{2\pi r p}{h} = n$$

which leads to the formula

$$2\pi r p = nh$$

Since $2\pi r p$ is the total action in one turn around the orbit, the formula is identical with Bohr's condition for allowed orbits! To summarize the conclusion: if an electron is described as a wave, with its wavelength determined by the momentum, then the standing waves produced by an electron circling a nucleus have the same momentum values as those for Bohr's orbits. As a final topper, de Broglie was able to show that the elliptical orbits later added to Bohr's theory were also standing waves.

De Broglie's thesis was a hot potato for the Faculty of Sciences of the University of Paris. A thesis, while expected to be original, isn't often *that* original. A solid contribution to an established topic is more common, even for the most brilliant student. And here was a convert from the humanities, writing on a theory not yet well known in France, promulgating an outrageous idea. De Broglie couldn't explain how the electron, which so far had behaved like a particle, could at the same time be a wave. In his own words, de Broglie characterized his theory as "a formal

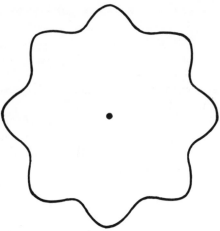

Figure 16-2 De Broglie's picture of a Bohr orbit.

scheme whose physical content is not yet determined." And aside from the curious coincidence of explaining the Bohr formula, itself still a suspect achievement, no experimental confirmation was offered.

But there were de Broglie's family connections to consider. His professors could hardly dare to insult a man who was at one and the same time a prince and the brother of an eminent colleague. Yet imagine the embarrassment if they were to award the doctorate for a crackpot thesis! The problem was resolved by de Broglie's sponsor, the eminent theorist Paul Langevin. He brought the young prince's work before the eyes of Einstein. The response was encouraging: "It may look crazy, but it really is sound!" Now matters were working to de Broglie's advantage. In Germany, home of the quantum theory, none other than the great Einstein was promoting his ideas. The story has an appropriately fairy-tale ending: the bold prince became the first (and remains the only) physicist ever to receive a Nobel prize for his doctoral thesis.

HELP FROM AMERICA

If the direct experimental proof that would have made his Parisian teachers breathe easier was lacking, it was not long in coming. Indeed, it had been there all along, though its significance had been overlooked. The scattering of electrons by matter had been studied for some time, and there were many features suggestive of a wave rather than a particle.

The best data had been taken by an American, Clinton Davisson. Like most United States physicists of his day, Davisson was a practical-minded man. His studies of the reflection of electrons off metals, conducted at the Bell Telephone Laboratory, were motivated by a desire to better understand the operation of vacuum tubes. When he found that electrons were reflected well at some angles and poorly at others, he regarded the result as a curiosity. He had barely heard of the quantum theory and did not have the benefit of an Einstein close at hand, filling his ear with wondrous tales of de Broglie's wild idea. When he learned that two German physicists, James Franck and Walter Elsasser, were touting his results as a proof of the wave nature of the electron, he was decidedly skeptical.

The Franck-Elsasser interpretation of Davisson's result is illustrated in Fig. 16-3. If the electron were a wave, only at certain angles would the portions of the wave scattered from different atoms interfere constructively. The most suggestive feature of the data was that the angles for strong reflection depended on the velocity of the electrons. This was easy to account for if de Broglie were right; it merely resulted from the changing wavelength of the electron. Calculating the wavelengths of the electrons from the angles reported by Davisson, they found agreement with de Broglie's formula.

Once again, the value of publishing raw experimental data had been demonstrated. Davisson, by reasons of education, motivation, and geographical isolation, could hardly have been expected to understand the full significance of his results. The availability of the data to the more cosmopolitan German physicists carried the day for de Broglie's crazy idea.

For two years Davisson continued his work, more bewildered than pleased by the furor it was creating in Europe. Finally, a trip to a

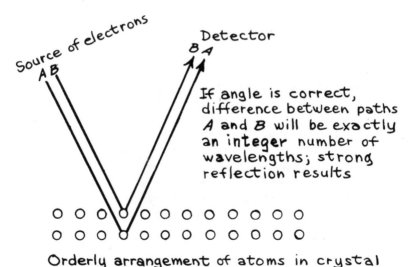

Figure 16-3 *Wave effects in the scattering of electrons off metals.*

scientific meeting in England helped Davisson to see the light. An able worker, he returned to the United States to attack the problem with enthusiasm. His final results were so conclusive as to leave no doubt. Davisson followed de Broglie into the august ranks of Nobel laureates.

A WAVE EQUATION

At this point, the reader is probably perplexed by the question of what all this wave business has to do with reality. How can a particle be a wave? What does the peculiar pattern in Fig. 16-2 have to do with an electron traveling in an orbit? This consternation is historically appropriate: de Broglie and his contemporaries were every bit as puzzled by this problem. But one thing was clear to them: no great light could be shed on the subject until they obtained a better picture of what the wave looked like and how it behaved. Perhaps then the connection between the wave and the particle would become clearer. The problem of the connection between the wave and the particle was the crucial issue in the development of the quantum theory.

One thing was obvious to all concerned: the simple picture in Fig. 16-2 was a bit *too* simple. True, some of the outer orbits, many wavelengths long, might look very much like that. The wave would be confined to a narrow ring. But the inner orbits must be very different. Because of the proximity of the nucleus, the force on the electron would vary rapidly with distance, and so would its speed. The corresponding wave would have a wavelength that varied from place to place; distorted in this way, its standing wave pattern would look very different from that shown in Fig. 16-2. A more fully developed theory was called for, one that would take into account the effect of a force on an object that was part particle and part wave.

The answer came in a matter of months from Erwin Schrödinger, a professor at Zurich. A true product of his time, Schrödinger had supplemented his Viennese education with short tours at Jena, Breslau, and Stuttgart. His thorough knowledge of mathematics convinced him that what was needed was a single, definitive equation whose solutions would describe the de Broglie wave in all circumstances. He quickly found it. It is reproduced below simply for the historical record; its mysterious symbols will make little sense to anyone with less than 2 years' training in the calculus. It is known technically as a *partial differential equation* and packs a remarkable amount of thought into a few symbols. The reader is advised to give it short attention and pass on to its "English translation,"

$$\left(-\frac{\hbar^2}{2m} \nabla^2 + V \right) \psi = i\hbar \, \frac{\partial \psi}{\partial t}$$

The symbol ψ is a mathematical function giving the strength of the de Broglie wave at various positions in space. In order to satisfy the Schrödinger equation, the wave must have the following properties:

1. The wave must satisfy the de Broglie rule $p = hk$.

2. Any force present is taken into account by means of the potential energy V it produces.

3. The wave obeys conservation of energy, in its classical (non-relativistic) form.

The mathematical prescription for finding the wave is thus very straightforward. At any given position in space, one subtracts from the total energy the potential energy at this spot, obtaining the kinetic energy. From the kinetic energy one can calculate the velocity and therefore the momentum, leading finally to the wavelength. Knowing the wavelength everywhere, one knows exactly what the wave looks like.

An example of this is provided for a very simple case in Fig. 16-4. This is the case of an electron passing through a thin sheet of metal. Here the force is quite simple. When the electron reaches the surface of the metal, it is attracted to the atoms on the surface. Hence, it speeds up. Its kinetic energy is increased and thus its momentum as well. Therefore, its wavelength becomes shorter. Once inside the metal, the electron experiences no further force; it is equally attracted in all directions, since it is surrounded by atoms. At its exit, it once again is attracted by surface atoms; the effect is to slow it down to its original speed. Viewed in terms of the conservation of energy, potential energy is converted into kinetic energy as the electron enters the metal sheet, and kinetic back into potential as it leaves.

The hydrogen atom was a far tougher nut to crack. Here the force

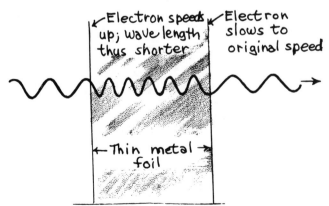

Figure 16-4 Wave picture of an electron passing through thin metal foil.

is present everywhere and varies with distance from the nucleus. Fortunately, the problem had already been solved, at least in part, by nineteenth-century mathematicians as a purely abstract mathematical exercise. Drawing upon these solutions, which bore such esoteric names as *confluent hypergeometric functions* and *spherical harmonics* (the latter term has always struck the author as a bit poetic), Schrödinger was able to give an exact picture of the standing electron waves that replace Bohr's orbits. Several of these corresponding to some of the smaller Bohr orbits, are depicted in Fig. 16-5. Where the pattern is brightest, the wave has greatest amplitude. Aside from being about the same size and corresponding to the same energy, they little resemble Bohr's circles.

Here at last was a complete theory, free from the ad hoc postulates of Bohr. The Schrödinger equation was the first quantum law that could pretend to the sort of generality possessed by Newton's laws. Any force, any situation whatsoever was covered by it, without any need for additional assumptions or arbitrary rules. Atoms turn out to have definite electron energies as a result of a phenomenon similar to that which makes musical instruments sound definite tones. It also eliminated the dual role of Planck's constant in the Bohr theory, for the orbit rule became itself a wave phenomenon, and thus both $E = h\nu$ and the orbit rule used Planck's constant to connect a wave phenomenon and a mechanical one.

The Schrödinger picture of the hydrogen atom did more than remove the arbitrariness of the Bohr orbits. It also eliminated from the quantum theory the even more vexing quantum jump from one orbit to another. The transition from one state (the word *orbit* hardly seems appropriate) to another, as described by Schrödinger, was a quite orderly process. Viewed in terms of the patterns shown in Fig. 16-5, it represents a kind of motion picture "dissolve" from one pattern to another. One wave pattern gradually fades out, while the new state fades in. During this time, light is being continuously emitted. The word "gradually" must, however, be interpreted in the context of the strange time scale of the atomic physicist: the whole process takes, typically, somewhere around 10^{-8} s.

The development of the Schrödinger–de Broglie wave theory did throw some light on the significance of Planck's celebrated constant h. Clearly, on the submicroscopic level of the atom, the concepts of wave and particle, carried over from the macroscopic world of our daily experience, begin to break down. At the same time, it becomes as meaningful to speak of *wavelength* and *frequency* as somehow related to *momentum* and *energy*. The constant h merely takes care of the quantitative aspects of going over from the wave language to the particle language. In this sense, it occupies a role similar to that of the velocity of light in the formula $E = mc^2$: it is no more than a conversion factor.

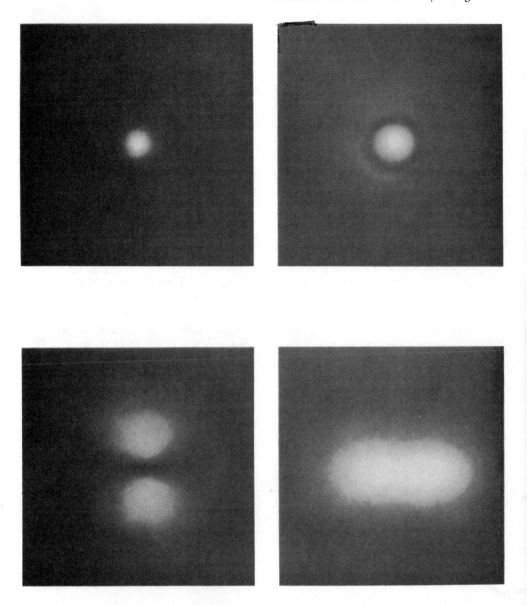

Figure 16-5 Examples of Schrödinger wave solutions for the electron in the hydrogen atom. The upper left picture corresponds to Bohr's original ground state; the others correspond to higher orbits. (Reproduced by permission from J. Orear, Fundamental Physics, 2d ed., John Wiley & Sons, Inc., New York, 1967.)

WHERE IS THE ELECTRON?

To Einstein and Planck, disturbed as they were by the peculiarities in the theory they had jointly fathered, Schrödinger appeared as a savior. At last the accursed quantum jump had been replaced by a beautiful shifting wave pattern. But at Copenhagen and Göttingen, things were seen in a very different light. Here the young physicists learned to live with, and even glory in, the discontinuities of the new physics. They were quick to point out that Schrödinger's theory left open the question of the connection between his waves and the obvious particle characteristics of an electron. Was one to believe that an electron actually smeared out into the sort of patterns shown in Fig. 16-5? There was a wealth of evidence that the electron was a perfectly good particle, probably smaller than an atomic nucleus. They were confident that the connection between Schrödinger's wave and the particle must in itself have some of the discontinuous features of the earlier quantum theory.

From their point of view, the real tip-off lay in Schrödinger's description of the very simplest situation in mechanics, the behavior of a free particle, one moving at constant velocity free from the influence of any force. In this situation, Schrödinger's equation gave a most embarrassing result: a free particle is described as a wave packet, as in Fig. 16-6, a bundle of waves confined to a small region in space. But a wave is not a particle, and this packet refused to stay small. Like the wake of a

Wave "packet" representing free electron

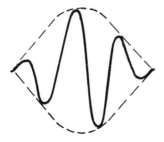

Same packet a very short time later

Figure 16-6

boat, which spreads out from being a single sharp bow wave to a whole train of waves, the wave packet would spread out and spread fast. If initially confined to a space the size of an atomic nucleus, it would spread in less than a millionth of a second to fill a space the size of the Pentagon!

While Schrödinger's imagination might accept an electron the size of a Bohr orbit, he could hardly reconcile himself to an electron as big as the Pentagon. The problem clearly had to be resolved. And its resolution touched off the greatest controversy in physics since Galileo liberated the science from theological disputes.

CHAPTER SEVENTEEN

Does God Play Dice?

At the final stage you tell me that this multi-colored universe can be reduced to the atom and that the atom itself can be reduced to the electron. All this is good and I wait for you to continue. But you tell me of an invisible planetary system where electrons gravitate around a nucleus. You explain this to me with an image. I realize then that you have been reduced to poetry: I shall never know. Have I the time to become indignant? You have changed theories. So that science that was to teach me everything ends up in a hypothesis, that lucidity founders in metaphor, that uncertainty is resolved in a work of art.

—ALBERT CAMUS, THE MYTH OF SISYPHUS

*I*F THE BOHR THEORY was the final break with classical physics, the interpretation of Schrödinger's wave was the break with the whole picture of what a physical law should be like. As with relativity, the "new" quantum theory based on waves is not so much obscure as strange; the statements seem simple enough, but their implications are unacceptable to our common sense. And this time what must be swallowed is even worse than a readjustment of such basic concepts as space and time. What the quantum theory seems to challenge is nothing less than the whole concept of continuity in nature. Words like *cause* and *effect* appear to lose their meaning. One speaks of motion but no longer dares to imagine a continuous path. The very functioning of reality on its ultimate level seems reduced to a cosmic dice game, everything subject to the whims of chance.

The formulation of this viewpoint took little more than a year after the publication of Schrödinger's theory and consisted of three main steps. First, Max Born gave a statistical meaning to Schrödinger's wave, connecting it to reality by pure chance. Then Werner Heisenberg showed that Born's interpretation could be regarded as the consequence of an irreducible interference of the observer with the system being observed. Finally, Niels Bohr interpreted this unprecedented situation as the denial that the quantum description of the microworld corresponded to "reality" in the traditional sense of the word. This chapter will outline and explain the first two developments.

MAX BORN'S DICE GAME

The problem facing the quantum theorists in 1926 was without parallel in the history of physics. For more than half a century, physicists had speculated with increasing confidence about the connection between unseen atoms and the readings on their laboratory instruments. Building invisible models became a fine art, and they could form mental images of what the atoms were doing, use reasoning based on these images to arrive at a formula for the things they could measure, and compare the predictions with the results. The example of Rutherford's experiment is an ideal one. The simple mental picture of a nuclear atom led to a formula for the angular variation of alpha-particle scattering, and the formula fit the measurements. One had no need to actually see the nucleus; the inference was certain enough.

But now the physicists had in Schrödinger's equation a formula that worked beautifully, yet one which gave no hint of how to visualize the reality that lay behind it. It would have been so simple if one could merely, as Schrödinger did at first, accept that the electron was a wave. Still, many experiments had shown the electron to be a particle: its mass and charge were confined to a small region, smaller than an atomic nucleus. One could even follow the path of a single electron in a photograph of a Wilson cloud chamber, and it certainly did not seem to expand like Schrödinger's wave packet. Yet the quantitative details of the behavior of electrons, both inside and outside atoms, were obtained by treating them as waves. And the connection between the wave and the particle, now that Schrödinger had painted the wave picture in full detail, was if anything less clear than it had been when de Broglie first proposed the duality. How could such an unruly wave, one that refused to remain confined to a small region, have anything to do with a particle? What mental picture was the physicist to keep in mind when visualizing subatomic phenomena?

Schrödinger. (Photo by Ullstein, courtesy of American Institute of Physics.)

In a sense, the scientific part of the job had already been done: the things one could measure, spectral lines and such, were predicted to the satisfaction of all concerned. Many physicists were reluctant to press farther, for since before the turn of the century strict empiricism had been the dominant philosophical view among natural scientists. The creed was, "Stick to your equations and instruments." The task of science was solely to find connections between measurements. To inquire into the reality that lay behind these connections, to ponder such hazy concepts as cause and effect, was regarded as both futile and unscientific, possibly no more than an exercise in language. Metaphysics was unfashionable, and the question of what an atom was really like seemed patently metaphysical.

Nonetheless, Schrödinger's wave seemed a permanent and ineradicable feature of the theory, and few physicists would be comfortable until they had some way to reconcile it with their intuition. What were the observable consequences, if any, of the fact that the wave was strong in one place and weak in another? There had to be some sort of answer.

Once again the answer came from Göttingen, from Max Born, director of the theoretical institute. In retrospect, Born described his motivations: "My Institute and that of James Franck were housed in the same building. . . . Every experiment by Franck and his assistants on electron collisions . . . appeared to me as a new proof of the corpuscular nature of the electron." Thus, he felt Schrödinger's electron "cloud" had to go. In the spring of 1926, a few short months after Schrödinger's celebrated papers, he proposed that Schrödinger's wave must be a *probability* wave; to be exact, *the square of the wave amplitude at any point in space gives the probability of finding the electron at that point.*

Under this interpretation, Schrödinger's cloudlike patterns take on a peculiar significance. They don't tell the physicist where the electron *is* at any given moment but merely where it is *likely to be*. An individual measurement of the position of the electron in, for example, the first Bohr state, can be as precise as the method used to measure it will allow. If one repeats the measurement many hundreds of times and plots the results in the form of dots on a picture, as in Fig. 17-1, the resulting pattern of dots will come to resemble the wave patterns shown in the preceding chapter. But no *individual* measurement can be predicted with any greater precision than to say it will fall somewhere in the cloud. Schrödinger's cloud becomes, in a sense, a cloud *in the human mind;* it reflects our lack of precise knowledge of where to find the electron. One way to say this is to describe it as an *information wave;* the exact significance of this remark will become apparent after we discuss Heisenberg's uncertainty relations, the next step in the interpretation of quantum mechanics.

Here was the quantum theorists' most daring innovation: determinism itself had been abandoned. No longer could a physicist measure

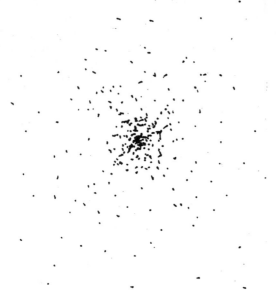

Figure 17-1 Result of repeated measurements of the position of an electron in the first Bohr orbit.

the position of an electron and use these measurements to predict with precision its future position. The wave itself remains perfectly continuous and determinate, but this is of no consequence, for the wave has only a random connection with observable reality: one can predict the average of a large number of repeated measurements, but the result of an individual measurement must forever remain a surprise.

Probabilistic laws were no novelty in physics—they had been present ever since physicists began to speculate about the motion of atoms. In the kinetic theory of gases, one always spoke of average speeds of atoms, average distances between collisions, without trying to trace the motions of the individual atoms in detail. This probabilistic character merely reflected practical ignorance; it was an acknowledgment of the impossibility of coping precisely with the motion of 10^{23} incredibly tiny atoms as individual objects. No one doubted that the details of the motion were subject to Newton's laws, and if one were given the staggering amount of information required, one could exactly describe the future motion of every atom. What Born seemed to be saying was that there was no way whatsoever to predict the precise future position of even one isolated atom.

UNCERTAINTY

One of Born's younger Göttingen colleagues, groping for a deeper understanding of his mentor's work, was struck by the probability-ignorance connection. Werner Heisenberg realized that the new quantum theory still retained much of the structure of Newton's mechanics. Through the wave-particle duality, it still permitted a description of motion in terms of ordinary concepts like position and momentum. In fact, Heisenberg himself had arrived at his own form of the quantum theory, one that gave many of the predictions of Schrödinger's, without ever mentioning waves. It was hard to see why, after one once knew precisely the position and velocity of a particle, its future could not be determined exactly. Did Born's interpretation imply that something would disturb it, deflect it somehow from the path assigned it by Newton's laws? Thinking in this vein, he had the key insight into the origins of the indeterminacy at the atomic level. He saw that this indeterminacy was indeed the result of ignorance, not merely a practical ignorance, but one of an inherent and unavoidable kind. Its source was *the disturbance of an object by the act of observing it.*

Classical physics had always removed the observer from the phenomenon observed. The flight of a ball is in no way altered by the presence or absence of a witness. It had been recognized that no measurement could be made without in some way disturbing things; even the light falling on the ball alters its path imperceptibly. But it was always assumed that careful experimental procedures could reduce the disturbance to a minimum, or at least enable one to correct for the disturbance. For example, a cold thermometer bulb will cause a small drop in the temperature of a beaker of hot water; but if one knows the original temperature and weight of the bulb, one can calculate the size of the effect and thus obtain the true temperature of the water before the insertion of the thermometer.

But quantum mechanics had changed all that, at least on the atomic scale. No longer could one arbitrarily reduce the disturbance caused by a measurement. At least one quantum of energy must be used to make an observation; and Heisenberg saw that furthermore the effect of this quantum was to disturb the object observed in a quite random and uncontrollable way, so that one could not correct for the disturbance! In quantitative terms, this is expressed in the statement that one cannot simultaneously measure the position and momentum of an object to any desired accuracy. No matter how good the instruments used or how careful the procedures, there had to be an irreducible error in at least one of the measurements. The quantitative statement of this rule is known as the *uncertainty principle;* it takes the form

$$\Delta p \ \Delta x \geq \textstyle\frac{1}{2}\hbar \tag{17-1}$$

Δp and Δx are the errors in measurement of position and momentum, respectively. What the law says is that one has to make a choice: if one measures position accurately, that is to say, if Δx is small, then Δp must be large to make the product of the two larger than $\frac{1}{2}\hbar$. The symbol $\geq$

("greater than or equal to") implies that ℏ is the *best* one can do. In a perfect instrument the product of the errors will be equal to ℏ. Any defect in the measuring procedure will make things worse, and the product of the errors will be even greater.

Any error in measurement of momentum, since mass is usually well determined, will lead to an error in velocity. And it is precisely the simultaneous knowledge of position and velocity that is essential to knowing where an object will be in the future. If we know how fast an object is going but have a poor idea of where it now is, we are just as badly off or worse off when it comes to predicting where it will be at some future time.

Heisenberg derived his law on quite general and abstract grounds. Its true significance becomes apparent only when one shows how it enters into any specific measuring process. This is another gedanken experiment game, like those we played in the development of relativity. The law works differently in every imaginable measuring process, but it is always there.

As just one example, consider using a microscope to measure the position of a stationary electron. The situation is illustrated in Fig. 17-2. To reduce the disturbance of the electron to a minimum, we use only

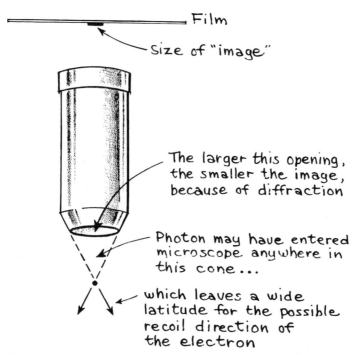

Film

Size of "image"

The larger this opening, the smaller the image, because of diffraction

Photon may have entered microscope anywhere in this cone...

which leaves a wide latitude for the possible recoil direction of the electron

Figure 17-2 *Uncertainty in observation of an electron by means of a microscope.*

one photon of light. The error in position comes about because of the wavelength of the light. The image formed by the microscope will give the electron a "size" of about 1 wavelength. What does *image* mean when we are dealing with a single photon? It means that there is a certain region in which the photon might strike the photographic plate. The actual spot on the film will be as small as the film quality permits; but if the measurement is repeated, the spot will not necessarily be in the same place. No matter how careful we are to place the electron in the same place, we cannot reduce the size of the spot when we make many repeated measurements. The situation is similar to the picture of a Bohr orbit in Fig. 17-1. This illustrates what (at least in this case) Heisenberg means by an *error*. It is not a question of the precision of the measuring instrument; a needle on a dial can be, at least in principle, as accurate as the instrument maker's skill permits. Or, as in this case, the size of a spot on the film depends on the film quality. The real problem is that of *repeatability*—a repetition of the experiment under conditions the experimenter has done everything possible to make identical will give a spot in a different place on the film.

Now comes the question of the momentum uncertainty, which in this case arises in a quite different way. When the photon rebounds off the electron, the electron must recoil. We have no way of knowing the *exact* direction of the photon when it entered the microscope lens, which must be, for reasons which will be explained below, quite wide. In this case, we can compute the *speed* of the electron after its collision with the photon to considerable accuracy; but if we don't know the *direction* it is heading, we have no idea where it will go.

In this and most other examples, the choice between accuracy in momentum and position reduces to a choice of wavelength. A long-wavelength photon has a low momentum, and thus the momentum uncertainty is reduced; but the position is less accurately known. A short-wavelength photon permits a more accurate position measurement, but gives the electron a bigger kick.

At this point it is to be hoped that your mind is searching for a way out of this mess; why not narrow down the lens of the microscope, so we know better where the photon went and therefore the direction of the electron recoil? This suggestion, however, ignores the wave character of light, which the existence of photons in no way negates. A wave passing through a narrow hole is spread out by diffraction; the smaller the hole, the larger the effect. In terms of the photon, just as the de Broglie wave represents the probability of finding an electron, the intensity of the light wave gives the probability of finding a photon. Narrowing the lens will widen the region of the photographic plate in which the photon may hit. The uncertainty principle is inexorable; any move we make that lets us determine the momentum more accurately must turn out to sacrifice accuracy in position. If the measuring instrument is perfect, the gain in accuracy of one variable is exactly offset by a loss in precision in the other.

Given that Planck's constant is a quantum of action, it should not surprise you that the uncertainty relation also holds for the pair *energy* and *time*. In this case, its significance is that *in order to have a well-defined energy, a physical state must last for a long time.* The energy of a short-lived excited atomic state is slightly "smeared out," as revealed by a good spectrograph. The lines are *not perfectly sharp;* each has a natural width that depends on how long the state it came from survived.

Heisenberg was by no means unaware of the ideological implications of his work. He was a member of the Youth Movement (*Jugendbewegung*), which strove for a romantic "renewal in body and spirit." In the pages of its journal Heisenberg proudly proclaimed the demise of determinism.

THE "OLD MEN" WON'T BUY IT

The shock and outrage of Einstein, Planck, de Broglie, and Schrödinger at these arguments were predictable. They had seen order and continuity restored to the microworld, only to see it snatched away a few short months later by Born and Heisenberg. Einstein in particular simply refused at first to accept the validity of the uncertainty principle, a position summarized in his celebrated remark that "God does not play dice!" His first reaction was to search for counterexamples, gedanken measuring procedures that would be exempt from the principle. But these proved as futile as the similar attempts made earlier against his own relativity theory. The others gradually and reluctantly came to accept the new view. A celebrated visit by Schrödinger to Bohr's institute in the fall of 1926 proved the turning point. After days of debate lasting well into the night, he finally conceded defeat with the outburst: "If one has to stick to this damned quantum jumping, then I regret ever having gotten involved in this thing!" He had started out to eliminate discontinuity from the quantum theory and had seen it instead enthroned in the very heart of the theory, in a way that drastically altered the physicist's traditional "gut feeling" for reality.

WHY THE WAVE EXPANDS

Armed with the uncertainty principle, we can now interpret Schrödinger's expanding wave packet and obtain some insight as to the significance of the wave itself. The size of the wave packet, at the outset, represents the uncertainty in our knowledge of the position of the electron, the Δx. The spread of the wave packet with time arises from the momentum uncertainty; since we don't know exactly how fast the electron is going, the longer we wait, the less certain we are as to where it is. This reveals the information-wave character of the Schrödinger wave. It represents not the electron itself but *what we know about the electron at any given time.* A large spread means simply that we can't predict the electron's position with precision. The Schrödinger wave is determined not solely by the nature of the electron and forces on it; it also reflects the most recent observation of it.

If we make a number of successive observations, following it along its path, the behavior of the wave (Fig. 17-3) is most peculiar. At each successive observation, we start all over again with a new and smaller wave packet. This reflects the fact that just before the observation we weren't quite sure where we'd find the electron. It could have been anywhere in the large area covered by the Schrödinger wave. Once the electron is observed, we can narrow it down to a smaller region, of a size determined by the precision of our measuring process; but the value of

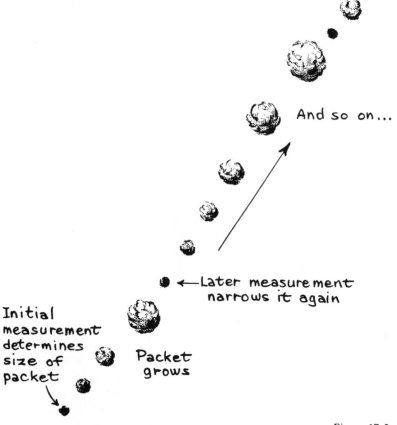

And so on...

←—Later measurement
    narrows it again

Initial
measurement
determines
size of
packet

Packet
grows

Figure 17-3

this information for future prediction is hampered by the uncertainty in momentum. Our ability to predict gradually deteriorates, until another measurement ties down the electron again.

The purely wave aspects of the spread of the wave packet can also be explained in terms of the uncertainty principle. The uncertainty in momentum corresponds, in the wave picture to a spread in wavelengths; the wave packet is a *superposition* of waves of varying wavelength. Since, in the case of the Schrödinger wave, waves of different length travel at different speeds, the packet will naturally spread out.

THE PROBLEM OF PREDICTION

Since the value of $\hbar$ is very small, the significance of the uncertainty principle depends crucially on the size of the objects involved. This is because it relates uncertainty in position to uncertainty in *momentum*, not velocity. It is the *velocity*, and not the momentum per se, that must be known in order to predict the future position. The more massive the object, the less important the uncertainty in momentum becomes, because it corresponds to a small uncertainty in velocity. This can be expressed mathematically in the form

$$\Delta p = m \ \Delta v \qquad (17\text{-}2)$$

Substituting in Eq. (17-1) and dividing by m to obtain an uncertainty relation between Δx and Δv gives

$$\Delta x \ \Delta v \geq \frac{\hbar}{m} \qquad \text{or} \qquad \Delta v = \frac{\hbar}{m \ \Delta x} \qquad (17\text{-}3)$$

(Since we are looking for a rough approximation, we have ignored the factor $\frac{1}{2}$). The larger the mass, the smaller the uncertainty. For objects visible to the naked eye, the mass is large, and also it takes a large position uncertainty to be noticeable. Thus, both m and Δx in Eq. (17-3) are much larger than they are for, say, an electron. The uncertainty in velocity becomes completely negligible, and we can use classical physics with full confidence.

This is illustrated quantitatively in Table 17-1. This table answers the question: How far in the future can I predict the position of an object to a certain desired level of accuracy? The answer is obtained in the following fashion. In the first column we choose a level of accuracy we consider appropriate to a given scale of phenomena. In the second, we choose the mass of an object that is to that scale. The third column gives the uncertainty in velocity of the object, as calculated from Eq. (17-3), given the mass and Δx from the first two columns. Finally, the fourth

TABLE 17-1

Scale and Δx, cm	Object and mass, g	Δv, cm/s	Prediction time limit, s	Remarks
Subatomic 10^{-9}	Electron 10^{-27}	10^9	10^{-18}	No continuity; classical physics essentially of no use
Atomic 10^{-8}	Atom 10^{-22}	10^3	10^{-11}	Can't describe whole history, but time long long enough to decribe collision with another atom very roughly
Biochemical 10^{-6}	Small protein molecule 10^{-18}	10^{-3}	10^{-3}	Long enough for nerve cell to fire; thus, short-term, functions can be described classically, though uncertainty very significant
Microscopic 10^{-4}	"Speck" 10^{-12}	10^{-11}	10^7	A dust mote barely visible in microscope; yet for almost all purposes completely classical
Macroscopic 10^{-3}	Pea 0.1	10^{-23}	10^{20} (3×10^{12} years)	Age of universe about 10^{10} years; can ignore uncertainty completely

column tells us how long it will be before the uncertainty in velocity will contribute as much to our lack of knowledge of where the object is as came from the original position error Δx.* After this time, the uncertainty due to our lack of knowledge of velocity will dominate and just get worse with time.

---

* Mathematically, $t = \Delta x/\Delta v$.

The lesson of the table is clear. Even if we take a fuzzy picture of the atom, i.e., only measure position to within about one-tenth the diameter of the first Bohr orbit, we cannot predict with much confidence beyond about 10^{-18} s—a fraction of the time it takes an electron to get around its orbit in the older Bohr atomic pictures! Yet on the scale of a small pea, we could wait 300 times the age of the universe before we would encounter a deviation of more than 0.001 cm (finer than a human hair) from Newton's laws.

Thus, the transition from the world of our everyday experience, where Newton's laws are for all practical purposes perfect, to the subatomic level, which is flagrantly "quantum," is a continuous one, governed by the size of h and the masses of the objects involved.

To summarize: the Schrödinger wave gives a perfectly continuous picture of the electron, or for the matter any other object. But the wave is spread out in space, reflecting the impossibility of precise knowledge of the electron's past and therefore of its future. And its relation to the future is determined by a toss of the dice. The situation is suggested in Fig. 17-3. And the whole mess can be thought of as the result of the irreducible interference of the observer with what he is observing.

THE ORBIT DISAPPEARS

One of the more disturbing consequences of the scheme is that it becomes quite difficult to think of the states of electrons in atoms as "orbits" any more, for we cannot observe them moving in their orbits. To do so would imply that we saw the electron at one position, then later at another, and so on mapping out a path.

But each of these observations again involves the exchange of at least one quantum with the electron. An electron in a hydrogen atom is not free to interact with any old quantum, for it is not free to assume any energy it pleases. The very *least* reaction it can have is to be knocked into some higher energy state. Thus, the orbit can only be observed once; the electron will be knocked clean out of it before we can see it again. We must be reconciled to the fact that if the electron follows an orderly path in the atom, it will never be within our power to map it out for any one particular electron. Under the circumstances, one has little choice but to admit that the word *orbit* is probably not a sensible one for describing an atomic state. Unfortunately, the theory offers no better one. We have learned to understand the structure of the atom only at the cost of giving up the sort of simple descriptive model that physicists are used to.

Even worse, though the electron has no path, it stubbornly retains all the other attributes of motion, momentum, velocity, and so on. In classical terms, to speak of motion without a path seems absurd. Yet this is what the quantum theory forces on us.

Thus, the quantum revolution is complete. It began with arbitrary rules that failed to describe all the details of subatomic motion. It ends with a complete theoretical framework that denies the possibility that man can ever have knowledge of these details! As we shall see in the next chapter, it led the Copenhagen school, Niels Bohr and his followers, to deny the very reality of the subatomic motions they originally set out to understand.

CHAPTER EIGHTEEN

Whatever Became of Reality?

The law of chaos is the law of ideas,
of improvisations and seasons of belief.

—WALLACE STEVENS, EXTRACTS FROM ADDRESSES
TO THE ACADEMY OF FINE IDEAS

THIS CHAPTER WILL illustrate by example the source of the perplexity that plagued physicists in the 1920s. On the surface, the uncertainty relations might be dismissed as a mere nuisance, the price we have to pay for being a very large bull in the tiny china shop of the atomic world. They certainly hamper our attempts to look at the submicroscopic world, but why should one imagine that they affect in any way the *reality* of that world? Even if one cannot observe an electron moving from place to place in an orderly orbit, is there any reason to deny the very existence of the orbit itself? In short, perhaps the uncertainty relations do no more than limit what can be *known* in a particular experiment.

But the example discussed in this chapter will show that it is not all that easy to separate the questions of knowledge and existence. They have a maddeningly intimate connection. Without this connection, Einstein would hardly have felt compelled to challenge the validity of the uncertainty relations themselves and would have dismissed them as a limit on the precision of atomic experiments, real enough, yet devoid of deeper implications.

YOUNG'S EXPERIMENT WITH ELECTRONS

The example chosen is that of Young's experiment, which we shall use as a gedanken experiment. It is hardly surprising that this particular experiment should be an apt one for our purposes. After all, it administered the *coup de grâce* to Newton's particle theory of light, and it seems only natural that a theory based on a wave-particle dualism would find it a tough nut to crack.

Young's experiment was introduced in Chap. 8 as a purely wave phenomenon. To emphasize the particle aspects, let us imagine it to be performed with electrons and a phosphorescent screen (such as that on a TV picture tube). Electrons are more tractable in another respect; they carry electric charge, and the forces exerted by this charge make it easy to observe when in flight. This is not true of photons, which can be created or absorbed but not observed in flight.

If we view the electron as a wave, the experiment is little changed from Young's original version. On the screen behind the slits, there are places where the distance to the slits is such that waves arrive crest to crest; here they are strong, and a bright band is obtained. At others, the waves meet trough to crest and cancel; this produces dark bands. The resulting pattern of light and dark bands is shown in Fig. 18-1. The only change introduced by quantum mechanics is that instead of a continuous glow, we will have a series of flashes at individual points on the screen. If there are enough of these in a short time, so that the eye doesn't see the

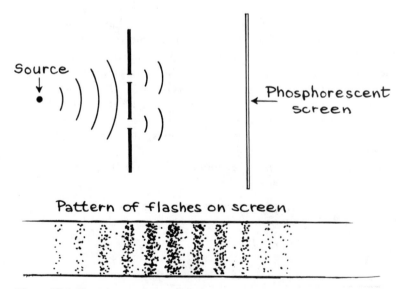

Figure 18-1 Young's experiment with electrons.

individual flashes, or if we wait long enough for a large number of them to pile up on a photographic plate, we will still obtain the pattern shown.

But from the particle point of view, the result is most perplexing. Surely each electron that produces its flash on the screen must pass through one or another of the holes. It is hard to imagine that the presence or absence of the other hole, the one it *didn't* go through, can affect the motion of the electron. If we were to close one hole, we would get a pattern without light and dark bands. Even conceding enough to the wave picture to admit that a particle is a wave packet and diffraction will cause the packet to fan out a bit after passing through a small hole, we still expect that a single slit will give the pattern shown in Fig. 18-2, and this proves to be the case. If we were to perform the experiment with each slit closed half the time, we would get a double exposure of Fig. 18-2, as shown in Fig. 18-3. Why don't we get this same pattern when *both* slits are open?

But a host of interference experiments with electrons have been successful. How can we account for this? Does each electron, as it passes through its hole, somehow "know" whether the other hole is open or closed, and accordingly "decide" whether to follow the Fig. 18-1 or the Fig. 18-2 pattern? Indeed, the electron's dilemma is even more complicated: the spacing of dark and light bands in Fig. 18-1 depends on how far apart the holes are, so the electron's behavior may even depend on how far it was from a hole it didn't go through!

The answer, of course, is that it does matter whether both holes are open or not, because *the wave describing each individual particle passes through both*. But at first sight this merely deepens the mystery. What then does this schizophrenic wave have to do with an electron, which measurement shows is to be a very small object? Could it have somehow split in two? It left its source and wound up all in one piece on a screen where it made one tiny flash as it came to rest. Surely it must have come through one hole or the other. Which one? We feel the wave picture must be giving us an incomplete description of what went on. If the theory will not answer our question, let us resort to experiment. Let us find out which hole each particle goes through, to see whether it indeed passes through one hole or the other, or whether it somehow splits and passes through both, like the wave.

The apparatus for this experiment is shown in Fig. 18-4. It is the

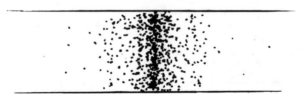

Figure 18-2 Pattern of flashes from a single slit.

Figure 18-3 Pattern of flashes from two slits never open at the same time.

same as before, except we now have some type of particle detector by each slit to tell us which one the electron went through. A loop of wire to sense the magnetic field of the electron as it passed would do this with a minimum of disturbance. Taking a very weak beam of particles, so that we can observe them one at a time, we find that each time there is a flash on the screen, one or the other particle detector will register—never both. The particle indeed passes through one hole or the other. But that is not all we discover. *The pattern shown in Fig. 18-1 has vanished!* We instead get the pattern shown in Fig. 18-3.

How can this be? The answer lies in the uncertainty relations. We have added another measurement into the experiment, and the effects of this measurement must be taken into account. To see the passage of an electron, we had to transmit at least one photon to the loop of wire. This changed its momentum, and therefore its wavelength, by an unknown amount, different for every electron. By the time it reaches the screen, the change in wavelength is enough to destroy the synchronization with the wave from the other slit, at which the detector remained silent.

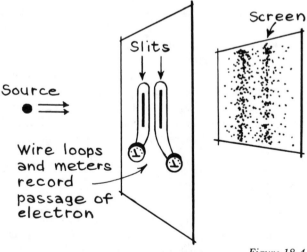

Figure 18-4

Figure 18-5 Partial interference pattern seen when an "unreliable" detector is used to tell which slit the electron passed through.

Once again, the uncertainty relations force us to make a choice. We can change to a larger loop of wire, which can detect a "soft" photon, one with low momentum and therefore a long wavelength. This disturbs the electron momentum less but leaves us more uncertain about its position. It turns out that if the size of the loop is small enough, and thus the photon is "hard" enough, to tell us the position of the electron to an accuracy much less than the slit separation, so that we can be certain which slit the electron passed through, the pattern is destroyed. If the error in position measurement is larger than the slit separation, the pattern is retained, but then one can't be really certain which slit the electron went through.

This result is not an either-or situation; intermediate cases exist. For example, we could use a detector whose accuracy was slightly smaller than the slit separation. This would never tell us for certain which slit the electron went through but might tell us, for example, that the odds were three to one that it went through one particular slit. A detector like this would only partially destroy the pattern, fuzz it out a bit, producing something intermediate between Figs. 18-1 and 18-3, as shown in Fig. 18-5.

This example illustrates vividly how the quantum theory faces a choice or at best a poor compromise between irreconcilable opposites. The wave and particle features cannot coexist in their pure form. One must choose between knowing which slit the electron went through (particle picture) and seeing an interference pattern (wave picture). One can never have both simultaneously. At best one can have an educated guess at where the electron went coexist with a smudged interference pattern. Niels Bohr called this characteristic of the theory *complementarity*, which he felt had deeper philosophical implications. We shall return to them later, after one last attempt to get around the problem.

CAN WE OUTWIT THE UNCERTAINTY RELATIONS?

The reader has every reason not to give up without at least one more attempt to thwart the uncertainty principle. After all, he is in pretty distinguished company, that of Einstein and others of comparable stature. The experiment shown in Fig. 18-4 isn't necessarily the cleverest one imaginable. In particular, it is beaten from the start because it disturbs the electron in flight. Why not let the electron hit the screen and *then* find out where it came from, by observing its direction? Then the new measurement is carried out *after* the interference pattern has already been formed unalterably. Uncertainty principle or no, surely one can't change things that have already happened!

The apparatus for this ingenious dodge is depicted in Fig. 18-6. Two screens are used, since one by itself won't reveal the direction of the electron that produced the flash. The first is made thin to permit the electron to make a flash, pass through, and make another flash on the second screen. The line joining the flashes surely will point to one slit or the other. We can have our cake and eat it too; so much for complementarity!

But again the uncertainty principle dashes our hopes. The electron can't make a flash without exchanging a quantum of energy with the first screen. This quantum gives the electron a kick, changing its direction. Of course we can use a very *small* quantum, to make the kick very small. But then the position of the electron when it passed through the first screen is poorly known. This means that the flash may come a great distance from where the electron went through. The result is often a flash where there should be a dark band; the interference pattern is unobservable. If we work out the quantitative details, we find the bind is the same one we were in with the arrangement shown in Fig. 18-4. If the kick Δp is small enough to still let us tell for sure which slit the electron passed through,

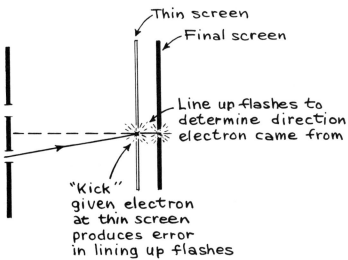

Figure 18-6 Apparatus for observing electron direction after pattern is produced.

Δx turns out to be larger than the spacing between bands. The result is a pattern like that in Fig. 18-3. Again, a compromise is possible: a Δp that still leaves a partial doubt as to which slit the electron went through will allow us to see an attenuated interference pattern, as in Fig. 18-5. What kind of pattern we see depends on how we choose the properties of our screen.

How frustrating! Quantum mechanics tells us there is a perfectly good interference pattern at the first screen. But if that screen is sharp enough to let us see the pattern of light and dark bands, it can't help destroying our knowledge of the electron's path. A conspiracy seems to be opposing our efforts.

Games of this sort can (and have) been played ever since 1927. An ingenious experimental setup is proposed that seems to get around the problem of complementarity, to permit a particle to exhibit its full wave and particle characteristics at the same time. So far, all such attempts have failed; it always turns out that the uncertainty relations have been ignored at some step of the argument. It hardly seems likely, after nearly 50 years, that this sort of frontal attack will bring down the edifice of the quantum theory.

THE COPENHAGEN INTERPRETATION

If we adopt a purely utilitarian view and make no demands on the quantum theory beyond the successful prediction of experimental results, the discussion could end here. But we undertake the study of nature not merely to master it but out of a deep-seated need to clothe our life experience in meaning. While younger minds were overcoming the formal mathematical problems of quantum mechanics, Niels Bohr struggled to find a deeper message behind the abstract symbols.

From his earliest days, Bohr had been influenced by the writings of Denmark's foremost modern philosopher, Sören Kierkegaard. A central tenet of the kierkegaardian view was the belief in an irreconcilable dualism between thought and reality. Rather than an impartial observer of reality, man was an active participant whose very effort to understand was part of the reality he falsely believed he was objectively contemplating. Reality understood is reality changed. The magnificent orderly systems that were the pride of classical physics struck Kierkegaard as narrow, symbolic interpretations of a tiny aspect of the full, rich panoply of reality.

Such dualistic views were popular with the rising empiricist school of philosophers. The American psychologist-philosopher William James viewed human thought as a conflict between the sensory images that constituted its "reality" and the rational thought processes that strove to organize and connect them.

Bohr found in these dualistic schemes a striking analogy to the wave-particle relation in the quantum theory. To a puzzled audience at an international conference in the fall of 1928, he announced his principle of complementarity. Wave and particle were irreconcilable, but both were necessary for the description of the subatomic world. On this level, complementarity was explored in the preceding section.

But to Bohr the wave-particle duality was merely a reflection, in the language of the old physics, of a deeper conflict between the goals of *description* and *causality*. One could describe the world, at any instant, to any desired accuracy, and produce a snapshot, so to speak, showing where everything was. Yet the uncertainty principle said that this snapshot could only be taken at the cost of forewearing any connection between it and future snapshots. The sharper the snapshot, the looser its causal ties with the future. The uncertainty principle is merely a symbolic statement, in the descriptive language carried over from the old physics, of this deeper conflict. We must choose some compromise between an orderly, causal world which we cannot even visualize and a sharp picture that reflects only the instant when it was taken. To visualize the atom in terms of concrete models is sheer nonsense. In Bohr's world there are no atoms; only observations. The atom is a creation of the human mind, to bring some order into the chaotic pattern of observation. The paradoxes and conflicts of the atomic world originate in the workings of the human consciousness.

Bohr's younger adherents eagerly welcomed this point of view. After all, they felt with pride that they had created a new physics. Now their mentor was telling them that not only had they overthrown the laws of classical physics, but the whole view of reality on which those laws were based. In its place was a new world view romantic enough to please a Spengler. The new physics had spawned an equally modern metaphysics.

In their elation, the proponents of this Copenhagen interpretation cast to the winds the traditional caution with which physicists approach philosophy. Young Johann von Neumann crowed that it was impossible to formulate the laws of physics without direct reference to the human consciousness and proved a theorem which purported to show that the Copenhagen view could never be separated from the quantum theory.

With the principle of complementarity, physics had come full circle from the days of Galileo. By removing man from nature, by making him a disinterested, nonparticipating observer, Galileo and his rationalist contemporaries had hoped to uncover a reality free from anthropomorphic bias. Now $3\frac{1}{2}$ centuries later, his intellectual descendants were denying that any such separation was possible.

THE DETERMINISTS STRIKE BACK

If Einstein and his followers were disturbed by the probability interpretation of the quantum theory, they were outraged by the deeper and more ambitious claims of the Copenhagen interpretation. Abandoning his unsuccessful direct attacks on the uncertainty principle, Einstein challenged the logical consistency of the theory as a whole. Though his arguments took many forms, they concentrated on the contraction of the wave packet that takes place when a particle's position is measured, which was described in the preceding chapter.

When an electron passes through a small hole, diffraction causes its wave packet to fan out rapidly. A short distance away, the packet covers an enormous fluorescent screen. When the electron flashes on the screen, the wave packet instantaneously contracts down to a tiny spot, for we know where the electron is.

The contradiction, in Einstein's view, lay in the word *instantaneously.* In less time than it takes for a signal to cross the screen, a physical event at one point (the flash) causes a disappearance of the wave. A physical influence has been transmitted at a speed exceeding that of light, in violation of relativity.

Yet to Bohr, Einstein's example was not a contradiction but a *proof* of the validity of his point of view. It merely showed that it was forever hopeless to ascribe any physical reality to the wave. The existence of a wave packet covering the entire screen merely reflected the possibility that the electron could strike the screen anywhere. An observer at the point of the flash knew instantly that the wave packet had vanished everywhere else; having seen the electron strike near him, he knew at once that it was no longer possible for it to strike elsewhere. Observers at other points on the screen would have to wait for the news to reach them, at the speed of light, before they realized the wave packet had shrunk. Since observers not in relative motion would disagree in the meantime as to the size of the wave packet, this was the final proof of Bohr's assertion that the wave packet represented not the electron itself but our *knowledge* of its position. It was Einstein's own search for a "real" interpretation of the wave packet, not Bohr's view of it as an information wave, that violated the principles of relativity.

Though he rephrased his argument many times in increasingly subtle form, Einstein ultimately was judged by most physicists to have lost the debate. The Copenhagen view quickly achieved the status of orthodoxy—at least among those heterodox physicists who even trouble their minds about such matters. Einstein regretfully conceded that Bohr's views were at least self-consistent. What he refused to concede was that they were the last word. Someday, he fervently hoped, a new discovery would restore order to the microworld.

Today Eugene Wigner, one of the last survivors of the golden era, is the principal standard bearer for the strict Copenhagen view, extending it to its purest form. He insists that the paradox of the quantum will only be resolved by penetrating the riddle of consciousness. He urges physicists to abet this process by lending their analytic methods to the science of psychology. David Bohm is the main spokesman for the opposite camp, a lonely seeker after determinism.

But time has a way of transforming the bizarre into the familiar, in science as well as in art. New generations of physicists have emerged, comfortable with the dualities of the quantum theory. Many suspect that the hidden message is not all *that* deep. It is simply that we should never have expected words born in the familiar world readily accessible to our senses, such as *particle* and *wave,* to perfectly describe the microcosm. The electron is what it is, and if the words we use to describe it seem full of paradox, *so much the worse for those words.* The equations have it pinned down neatly.

So the torch has been passed to the new generation, the quarrel left to the old. We are eternally grateful to them for landing us in the new world, but now we must get about the job of building a new physics in it. For the quest did not end with 1928. The quantum theory is no more than a scalpel to cut into the heart of the atom. It does not tell us what we will find there.

Let us close this book with a brief look at what is up among the people who ask *what the things the things that atoms are made of are made of.*

CHAPTER NINETEEN

The Dreams Stuff is Made Of

Three quarks for Muster Mark!
—JAMES JOYCE, Finnegans Wake

SINCE ITS BIRTH 2500 years ago in city-states of ancient Greece, no idea has shown quite the power to enchant the scientific mind as that of the atom. In the mid-1970s it burst forth on the world scene once again, filling the newspaper headlines with the mysterious word *quark*. So this final chapter will bring the story of physics up to the present moment, to discoveries made literally within weeks of this writing and to the author's own field of research.

A caveat is in order. When writing of events so recent, one always runs the risk that the words will be out of date even before they see print. But in this case, what has triumphed is a new and strikingly simple picture of matter on a scale 10 million times smaller than the atom. Though the details of this picture are still changing on a month-to-month basis, the basic outline is now firm and clear and likely to be with us for some time to come.

FAREWELL TO INNOCENCE

After any scientific revolution, a few loose ends always remain. To the builders of the new theory, they usually seem to be details that will be cleaned up in due course, and often this proves to be the case. But sometimes they prove the seeds of the next flowering, seeds that may take decades to germinate. So it was with the quantum theory of 1928. The major loose ends were two:

1. Schrödinger's equation was nonrelativistic.

2. The treatment of electromagnetism was not self-consistent; light was quantized in the form of photons, while the force inside the atom was as continuous as any that Newton had ever dealt with.

In the early 1930s, the British theorist Paul Dirac saw that both loose ends could be tied up neatly in a *quantum field theory.* But before this theory was complete, Europe's creative spree had come to a violent end. The Nazi takeover fragmented the German scientific community, sending many of its leading lights into forced or voluntary exile and leaving most of the rest demoralized.

Bohr's Institute took on a less frivolous air, as it became a way station for fleeing exiles. Bohr himself became increasingly preoccupied with the task of finding them jobs in safe countries.

And then, in the Christmas season of 1938, fission was discovered at the Kaiser Wilhelm Institute in Berlin. Within a matter of weeks, the wizards of the microworld knew that their days of happy innocence were at an end. A nuclear bomb might well be possible, and Nazi Germany had a head start. As if to underscore their apprehensions, a lid of military secrecy was clamped on a new wing at the Kaiser Wilhelm, and the Nazi state seized the uranium-rich tailings of the Czech radium mines. A grim race was on, and so far it had but one runner, a fanatic bent on world conquest. Refugee physicists pushed Great Britain and then the United States into the contest, but only after two precious years had been lost. The quantum would have to wait.

By the time the quantum physicists were free to return to their first love, the torch had been passed to America. Dirac's theory was completed in 1947 by two New Yorkers still in their twenties, Richard Feynman and Julian Schwinger.

QUANTIZING THE FIELD

Dick Feynman would have been in his element in the halcyon days of Bohr's Institute (he is in fact a student of Bohr's leading American disciple, John Archibald Wheeler). Fond of high jinks and high living, he has been known to jolt his mind out of a rut by working at a back table in a topless bar, inspired rather than distracted by the glare of stage lights and the blare of the jukebox. Schwinger, on the other hand, prefers Einsteinlike solitude. Working independently, they completed their theories within weeks of one another. As might be expected with two such contrasting figures, the theories looked on the surface so different that it took a considerable amount of labor to prove they were in fact the same.

The essence of Feynman's style is simplicity, so it is his version of the theory that we present in Fig. 19-1, which shows the quantum version of how the electromagnetic field transfers energy and momentum between two electrons. Newton's continuous force is replaced by a "package" transfer in the form of a photon. When the particles are far apart (and, by the standards of this theory, electrons in atoms are far apart), many small transfers occur and the force is almost continuous. But in a brief encounter of swift-moving electrons, the process is dominated by the exchange of a single photon. The force law is replaced by a formula that gives the *probability that any given amount of momentum will be transferred.*

This figure, known as a *Feynman diagram,* is more than a way to visualize the process. It contains an exact recipe for calculating the probability. More complex processes lead to more complex diagrams, containing more photons. Each line in the diagram and each vertex (intersection) contributes a specific term to the equation. The probability is the product of all these terms. Thought the calculations are tedious, once the right diagram has been drawn, they are so automatic that they can be turned over to a computer.

The terms contributed by the *vertices* are the most important and also the simplest for they determine the basic strength of the force. Each vertex contributes a constant proportional to the square of the electron's charge,

$$\alpha = \frac{e^2}{\hbar c} = \frac{1}{137}$$

which can be regarded as the *intrinsic probability that an electron will emit or absorb a photon.* Obviously, the higher the probability, the stronger the force. Such a number is called a *coupling constant.* It reduces the electron charge to a pure number, i.e., one which is independent of the choice of units.

The contribution from the *internal lines* of the diagram, those which *connect two vertices,* are somewhat more complicated. These terms determine how the probability depends on momentum transfer. The key to this is how quickly the process takes place. To see why, look at it from a reference frame in which the electron that emits the photon is at rest. Initially, the electron has only its rest energy. Afterward, kinetic energy and the photon energy are present. After the photon is absorbed, things are balanced out again, for the moving electron loses energy. The uncertainty relation says this is all right as long as it happens fast enough, since for a short enough time the energy uncertainty can cover for the

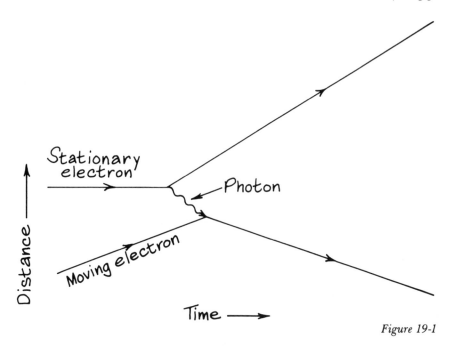

Figure 19-1

energy imbalance. The farther apart the electrons, the longer the photon must be in flight, so the transfer of large amounts of momentum and energy becomes improbable.

For each internal line, the uncertainty principle is invoked to produce a mathematical expression, or *function,* that gives the probability in terms of the momentum and energy carried by the particle.

Nonetheless, the field can extend to vast distances, for a photon, having no rest mass, can transmit as little energy as we please. Long-range electrical forces are transmitted by a continuous hail of photons of negligible energy.

But one of the more striking discoveries of the 1970s is that *the photon is not the only quantum of its field.* There are at least three more, and these others have a very considerable rest mass, more than 80 atomic mass units! These quanta have never been seen in the free, lightlike state because they are too heavy to produce with existing machinery. But the force they generate has been studied for many years.

Because of the huge rest energy of these quanta, the uncertainty relation allows them to be in transit for only short times. Thus the force they generate cannot extend to vast distances. In fact, it is restricted to encounters within 10^{-16} cm, 100 million times smaller than an atom. Within this range it is roughly as strong as that carried by photons, for its strength is fixed by the same coupling constant α. But since such close encounters are rare, in practice the force is very weak. For this reason, before it was recognized as a form of electromagnetism, it was given the name *weak interaction.*

The external lines, those which enter and leave the diagram, do not contribute to the probability. But because they represent free particles that can be observed for a long time and over great distances, so that their energy and momentum can be determined to high precision, they must conserve energy and momentum.

Thus quantum field theory reduces the calculation to a mechanical process, once the correct Feynman diagram has been drawn. Each vertex contributes a constant, each internal line a mathematical function, and the external lines provide connections between the momentum and energy of the various particles involved. What makes the computation tedious is that often more than one Feynman diagram can contribute to the same end result. The theorist must be careful to consider all diagrams that contribute a significant amount ot the probability.

So just in general relativity, the field concept that began modestly as a substitute for action at a distance materializes as *matter itself.*

MATTER AND ANTIMATTER

One of the more interesting predictions of quantum field theory, present in Dirac's original version, is that for each type of particle there must exist an *antiparticle,* opposite in electrical charge but equal in mass. Within a few years of the prediction, it was confirmed by the discovery of the *positron,* identical to the electron in all respects save that it carries a positive electric charge. It has since been confirmed for many other types of particle. Thus our world of matter is mirrored by a world of *antimatter* in which electrons are positive and protons negative.

A *fundamental particle* such as the electron can be created only if at the same time its own antiparticle is created. Similarly, it can be destroyed only if it encounters one of its own antiparticles. *Field quanta* such as the photon, however, can be freely created or destroyed. This provides a clear-cut distinction between field quanta and fundamental particles.

Figure 19-2 shows the Feynman diagram for the creation of an electron-positron pair when a photon with sufficient energy passes close to an atomic nucleus. The nucleus must be present because the photon uses up part of its energy to create the mass of the positron and electron. The pair thus has less momentum than the original photon, and the excess momentum must be carried away by the nucleus. When a positron meets an electron, they annihilate one another. Their rest energy goes into the creation of photons or other particles.

These rules are a severe embarrassment to cosmologists. If they

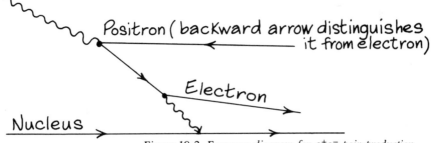

Figure 19-2 Feynman diagram for e⁺e⁻ pair production.

were obeyed when our universe was formed, then *where did all the antimatter go?* We know that our corner of creation, meaning our galaxy and its near neighbors, consists entirely of matter, except for an occasional stray antiparticle. Nowhere in the universe do we see the kind of titanic explosions that happen when large hunks of matter and antimatter meet. So if our universe does contain as much antimatter as matter, they are surely well segregated. Cosmologists are at a loss either to explain the process of segregation or to find a way of creating unequal amounts of matter and antimatter.

THE PROTON AND NEUTRON WON'T DO

Quantum field theory gives us a prescription for a picture of matter:

    1.   Everything is made of fundamental particles held together by field quanta.

    2.   Both are simple, pointlike objects.

The latter statement does not necessarily imply that electrons are *geometric points;* it is enough that they be so small that their size can be safely ignored in the theory. Measure the electron's charge and mass, and you know all you need to about it.

Where we go from here is obvious; the task of subatomic physics is to *find all the fundamental particles and field quanta,* and then we can close up this particular shop and move on to other problems. Easy enough to say, but it proved so hard to do that as recently as 1974 many physicists working in this area were quite prepared to chuck quantum field theory and look for some new approach.

In the heroic early days of the theory, it looked for a while as if the task was done, at least as far as the fundamental particles were concerned. The *proton,* or hydrogen nucleus, was an old familiar friend. In 1931 its neutral counterpart, the *neutron,* was discovered. At last physicists knew what nuclei were made of. All that remained was to find the field that holds them together.

There was one small cloud threatening the picnic: the proton and neutron are *not* pointlike. They are both about 10^{-13} cm in diameter. Though this is 100,000 times smaller than a typical atom, it is not much smaller than the spacing between them inside a nucleus. Quantum field theory just could not get to first base in the nucleus.

Then in the 1950s and 1960s came an unwelcome surprise; the proton and neutron are far from the end of the story. They are members of a vast family of particles, numbering in the hundreds. The rest are all short-lived, but otherwise they have an equal claim to the title *fundamental*. What now for the promise of simplicity that lured science into the game of atomism?

Furthermore, when the first fuzzy pictures of the inside of protons and neutrons emerged, they more nearly resembled Swiss watches than billiard balls. For example, the neutron is not really *free* of electric charge, it simply has *equal amounts of positive and negative charge*, arranged in a complex pattern. To have the "atom within the atom" messier than the atom itself was, by the stark rules of atomist aesthetics, ugly enough to turn men to stone. Would this endless Chinese puzzle of boxes within boxes never cease? There was only one way to find out; *build the machines to pry open the next box.*

THE ATOM SMASHERS

It sometimes strikes even the physicists who work with them as ironic that some of the largest machines mankind has ever built are used to look at the smallest things we know about. The 4-mi ring of Fermilab, west of Chicago, is clearly visible from well out in space. The blame can be put squarely on the Heisenberg uncertainty relation.

Peeking inside subatomic particles means looking for details smaller than 10^{-13} cm. This is tantamount to measuring distances this small. Heisenberg won't let you do this unless you transfer huge amounts of momentum. To transfer it, first you must have the momentum, protons and electrons with tens or hundreds of billions of electronvolts* of energy.

The relativistic rest energy (henceforth we shall simply use the word *mass,* for we are now in the realm where energy units are used for mass) of a proton is 938 million electronvolts, while that of an electron is about a half million. Thus these particles must be pushed to speeds within an eyelash of that of light, making them hundreds or thousands of times

---

* The electronvolt (eV), the most common energy unit in atomic and subatomic physics, is equivalent to 1.6×10^{-19} J.

heavier than at rest. To do this takes machines built on a heroic scale, of which there are only a handful in the entire world. The physicist who wishes to use one must be prepared to travel halfway around the globe to the laboratory where the experiment can be set up most conveniently and cheaply. Cheaply because monuments to curiosity simply cannot be funded on the scale common to defense and space research, in which national security and pride are presumed to be at stake. Curiosity has always run a poor second to pride where governments are concerned.

Beams of particles from *accelerators,* as these machines are called, smash into solid targets. There they collide with nuclei, converting much of their energy into mass in the form of new and often unstable particles. The debris is carefully scanned for clues to the inner structure of the particles. A picture emerges slowly, by combining the results of many experiments.

At first, the main job was simply to find and catalog all the types of particles, the sort of unpleasant but essential "dog work" that physicists would rather avoid. By 1964 the list was long enough for two theorists, Murray Gell-Mann and George Zweig, to hazard a guess at what they might be made of. Since at the time little was known of the inner workings of the particles, this feat was rather like reconstructing the parts list of an erector set from a list of the toys it can make. Thus was born the celebrated *quark model,* which was to ride to triumph a decade later.

Quark sounds more like the cry of a duck in distress than a serious scientific term. Such is the irreverent approach to scientific jargon that is the trademark of particle physicists and especially of Gell-Mann, who in his forties still retains some the whiz-kid style that earned him a Ph.D. from MIT at the remarkable age of 21. The source of this peculiar word was revealed at the head of this chapter.

THE QUARK THEORY

The history of the quark theory is such a tangled web that to avoid confusion we present the picture as it stood in October 1976, reserving the story of how it came to be for the next section. The theory can be reduced to the following statements:

1.   *Quarks are fundamental pointlike particles.* Their antiparticles are called *antiquarks.* We will use the symbols q and $\bar{q}$ to denote a quark and antiquark.

2.   *There are four kinds of quark* (the technical term is *flavors*). They differ solely in mass and electric charge; otherwise, they are identical. Whether more quark flavors remain to be discovered is an open question, but they would not change the theory in any essential way.

3. *The weak interaction can change the flavor of a quark.* This usually changes a heavier quark to a lighter one, but if energy is available, as in a collision, it can go the other way.

4. *Quarks are inseparable.* No quark has ever been seen all alone, only bound to other quarks or antiquarks. The field responsible for this unbreakable grip is not yet understood, but we have some tantalizing hints that will be discussed later.

5. *Quarks can combine in only two ways:* a quark with an antiquark ($q\bar{q}$), called a *meson,* or three quarks (qqq) or antiquarks ($\bar{q}\bar{q}\bar{q}$), called *baryons* and *antibaryons.* The proton and neutron are the lightest baryons and therefore the most stable. No meson is stable since as soon as the weak interaction has matched the flavors of quark and antiquark, they are free to annihilate.

6. *Like atoms, baryons and mesons have excited states.* The force binding quarks is so strong that these states are far apart in energy and hence have appreciably different masses. For this reason, the excited states were originally thought to be particles in their own right, which accounts for the proliferation of particles in the 1960s.

7. *The electron retains its status as a fundamental particle.* It belongs to a separate family, related to the quarks, called *leptons.* There are at present exactly as many flavors of leptons as of quarks. The weak interaction transforms leptons in exactly the way it transforms quarks. This is the main clue that the two families of fundamental particles are related. Leptons are immune to the force that binds quarks, just as electrically neutral objects are immune to electromagnetism.

Figure 19-3a gives the complete pedigrees of the quark and lepton families. The symbols *u, d, s, c* stand for the names of the flavors: *up, down, strange,* and *charmed* (the last two whimsical names refer not to any property of the particles themselves but to how physicists felt about them when they were first discovered).

As for the leptons, literally hundreds of experiments have shown that the sole distinction between the electron and the muon is in their mass, a situation that prompted I. I. Rabi to remark, "Who ordered that?" Whether the two kinds of neutrino have any mass to distinguish them is not known; we simply know they are lighter than electrons by a fair margin. We recognize them as distinct because different reactions ensue when they strike a nucleus.

All combinations of quarks and antiquarks are used to form mesons and baryons. The meson case is simple; since there are four quarks and four antiquarks, $4 \times 4 = 16$ different kinds are possible. The baryons are a bit more complicated, but by writing down all the combinations, as in Fig. 19-3b, we find there are 20. The rest of the 300 or so baryons and mesons are simply *excited states* of the basic combinations.

Leptons

Charge

	e	μ
−1	(0.51) electron	(106) muon
0	ν_p (0?) e neutrino	ν_μ (0?) μ neutrino

Quarks

	d	s
$-\frac{1}{3}$	(335) down	(540) strange
$+\frac{2}{3}$	μ (335) up	c ($\sim$1700) charmed

Figure 19-3a Fundamental particles. Masses in parentheses are in millions of electron-volts.

Another peculiar feature of Fig. 19-3a is the electric charges of the quarks, either $-\frac{1}{3}$ or $+\frac{2}{3}$ of the fundamental unit. Such fractional charges never appear in baryons, which always contain three quarks and thus can have only -1, 0, $+1$, or $+2$ units. Since antiquarks have charges $+\frac{1}{3}$ and $-\frac{2}{3}$, only charges of -1, 0, and $+1$ are possible for mesons.

Despite the rich variety of quark combinations *only two*, the proton and neutron, play any role in ordinary matter. The weak interaction quickly eliminates the baryons containing strange or charmed quarks; none last more than 10^{-10} s. Annihilation disposes of a meson in no more than 10^{-8} s. Such particles can be studied only during the brief instant when they flash by particle detectors at the accelerators where they are produced. While they thus seem to play no great roles in the cosmic scheme of things, they have been superb teachers in their short, fleeting lives; without them, we never would have solved the riddle of the atom within the atom.

Quark combinations

20 baryons

```
u u u
d d d
s s s
c c c
u u d  (proton)
u d d  (neutron)
u u s
u s s
u u c
u c c
d d s
d s s
d d c
d c c
s s c
s c c
u d s
u d c
u s c
d s c
```

16 mesons

```
u ū
u d̄
u s̄
u c̄
d ū
d d̄
d s̄
d c̄
s ū
s d̄
s s̄
s c̄
c ū
c d̄
c s̄
c c̄  (psi)
```

Figure 19-3b

DEMOCRITUS OR BOSCOVICH?

In order to understand why the quark theory needed a scientific minirevolution to gain final acceptance, it is important to realize that *some of its most attractive features, namely 1, 4, and 7 from the preceding section, were not part of the original 1964 theory.* In addition, the fourth ("charmed") quark had not been discovered at that time.

Feature 4, the *inseparability of quarks,* was especially important. It came as an experimental surprise, one which many physicists took as a blow fatal to the theory.

The hunt for a free, unattached quark, which lasted for a decade, ranks as one of the most frustrating chapters in the history of physics. The usual technique was to blast a beam of protons into a target and send the debris down an obstacle course of magnets that only a fractionally charged particle could successfully run. Other experimenters, aware that existing accelerators might not provide enough energy to overcome quark binding, turned to cosmic rays. These are very energetic protons that come down to earth from space. It is believed that they trace their origin to supernova explosions. If we simply wait long enough, a cosmic ray of almost any desired energy will eventually happen along.

Though there were a few false alarms, all these experiments, which numbered more than 50, failed to turn up a free quark. In their despair, many workers in this field concluded that quarks must be something else than particles. *What* else was anyone's guess.

Nonetheless, the theory was an unquestionable success. As hundreds of new baryons and mesons were found, not one appeared that could not be built up from quarks, and none of the quark combinations was missing.

This left particle physics in a most confused state. The proton, despite all the evidence for its internal structure, was a perfectly legitimate "atom" in the sense that it met the ancient test, inherited from Democritus, of *indivisibility*. At this point, Richard Feynman intervened in a decisive fashion. Though he had not taken part in shaping the original quark theory, he reminded his collegues that if it was to serve as the basis for a quantum field theory of matter, the real test of atomicity should be taken *not from Democritus but from Boscovich;* the crucial question was *not whether the proton was divisible* but *whether the quarks were pointlike!*

Feynman devised tests to apply to data on particle collisions that would reveal whether the particles contained pointlike parts. Particularly useful were collisions of electrons with protons and neutrons because in this case at least the electron was known to be pointlike, simplifying the analysis. By 1974 the results were clear; all of the electric charge inside protons and neutrons is confined to small regions at least 500 times smaller than the proton itself, and the charges are fractional, as predicted by the quark theory.

There was nothing terribly new about this approach; Feynman was simply treading the familiar trail blazed by Rutherford. The situation was complicated somewhat by the fact that the quarks are bound together by an unknown force, which made it hard to interpret the evidence unambiguously. Feynman's gift to physics consisted mainly in his dogged insistence on a simple picture and his deep faith that a quantum field theory was the simplest of all.

The trouble was that few physicists shared this faith. Most had their minds focused on the maddening refusal of quarks to come unglued. Until this could be understood, it seemed hard to believe that Feynman's simplistic approach might actually work.

The stalemate was resolved with dramatic swiftness by discovery of the $c\bar{c}$ meson, dubbed the *psi*, in November, 1974. With four quarks instead of three, the parallel with leptons was a compelling hint that they might really be fundamental. But that was not all; at last, an experimental handle by which to grasp the problem of quark binding fell in the laps of the physicists.

THE UNBREAKABLE STRING

On the long lonely nights at the accelerator laboratories particle physicists often lamented that Mother Nature had not seen fit to be as kind to them as she had been to Niels Bohr. Nowhere in the confusing tangle of quark states was a pattern as starkly simple as the orderly spectrum of hydrogen. Then suddenly, in the psi, this "oversight" was rectified.

For the c quark is so much heavier than its sisters that there is no mistaking the excited states of the psi. They form a clear, sharp grouping far heavier than the excited states of other mesons. By the spring of 1976, about 10 had been found. The pattern of *energy levels* they form is depicted in Fig. 19-4, along with the hydrogen-atom level scheme for comparison.

Bohr had worked with a well-understood force, the inverse-square attraction of electricity. This enabled him to use the energy-level scheme of hydrogen as a guide to his quantum rules. In 1976, with more general versions of these rules clearly established, it was possible to *reverse* Bohr's reasoning and *use the energy levels of the psi as a clue to the force* that locks the quarks in their unbreakable embrace.

The answer proved so ridiculously simple that a generation of physicists inured to complexity and confusion could scarcely believe their eyes. *It is a force that does not diminish with distance!* Furthermore, the force is immense; about 10 tons.!

A comparison of the psi energy levels with those of hydrogen reveals these basic features. The spacing of energy levels is the key to the strength of the force, for it shows how much work is required to move from one level to the next. Not only are the psi levels 100 million times farther apart, but the higher hydrogen levels crowd together, showing that the force diminishes far from the nucleus. This does not happen to the psi; the force holds its strength, undiminished, as quark and antiquark move apart. Inexorably, it must reel the wayward particles in.

If we retreat for a moment from quantum to classical field theory, we have a situation that Michael Faraday, with his incisive geometric intuition, would have seen through in an instant. If you want a force that does not diminish with distance, *bunch up all the lines of force into a tight bundle connecting the two particles.* Then they do not spread out in space, and the inverse-square law is "repealed." (Figure 19-5.)

In the quantum version, the bundle of lines of force becomes a "string" of field energy set to materialize in the form of photons (or, more often, new quark-antiquark pairs) as the wayward excited quarks are reeled in. Thus a highly excited particle disintegrates into a shower of mesons. The force is so strong that this takes only about 10^{-23} s.

Of course, one significant mystery remains, *the binding of baryons.* These peculiar three-quark combinations are a clear hint that the quark binding field has cuter tricks up its sleeve than a simple "opposites attract." Our past experience with field theories warns us to be patient in awaiting the solution to this riddle; the 16-year lapse between the Dirac and Feynman theories cannot be blamed *entirely* on World War I; a new field theory can be a herculean labor.

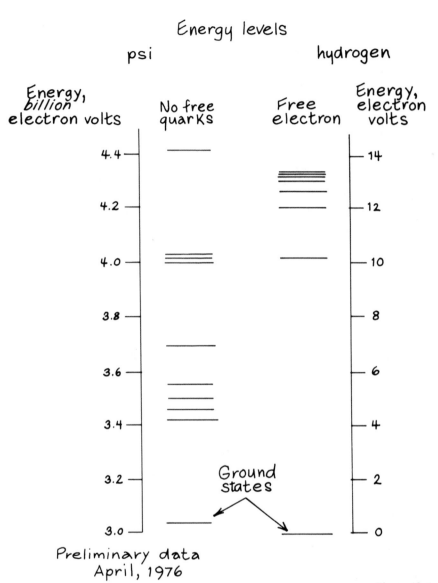

Energy levels

psi · hydrogen

Figure 19-4

Still, there are hopeful signs. Baryon binding has the same strength as that of mesons, and excited baryons produce the same meson cascade as they return to normal, leaving little doubt that the field is the same as that which binds mesons. Furthermore, the bunching of field lines can turn a weak force into a strong one, so it is possible to venture a vagrant hope that when the smoke clears, the field will prove to be good old electromagnetism in yet another guise.

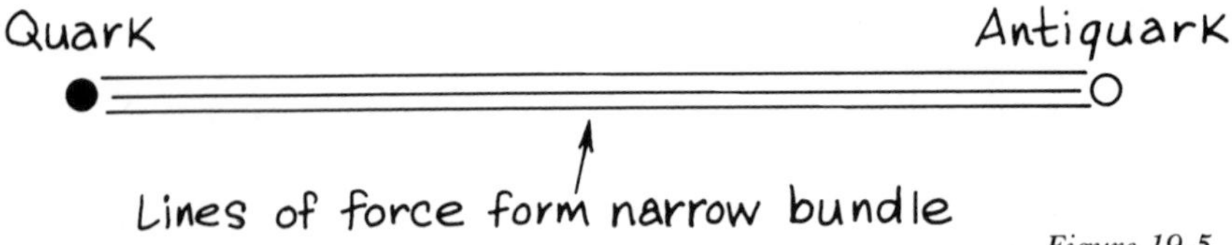

Quark   Antiquark

Lines of force form narrow bundle

Figure 19-5

But while awaiting the slow gestation of a *super unified field theory,* the curious types who choose to work on this frontier can no more resist the temptation to speculate than they can stop breathing. Some of the more novel speculations concern the connection of the microworld with gravity and the big bang. So, with the clear warning that nothing that follows has any but the most tenuous connection with observed fact, let us see what happens when we try to weave together the two main strands of twentieth-century physics.

THE QUANTUM FOAM

One of the obvious questions, in the wake of the triumph of the quark theory, is: How big is a point? Will quarks and leptons, on a scale smaller than our present 10^{-16} cm limit, reveal that they too have inner workings? Is there any end to this game?

A partial answer may be sought by combining Heisenberg's uncertainty relation with the mysterious black hole. Take a small region of space-time. The uncertainty relation says that we cannot be sure how much energy (and therefore mass) is contained therein. As we make the region smaller, the uncertainty grows. With it grows the size of a black hole of mass equal to the uncertainty, for this distance is proportional to mass. If one length grows as the other shrinks, eventually the two must cross; the Heisenberg rule informs us that we can never tell for certain whether this small a region is or is not part of our universe!

This limit, which has been christened the *Planck length,* is 10^{-33} cm. Its significance requires some thought. At the very least, it sets a limit to the meaningful size of any object. John Wheeler sees it as a *quantum foam;* it implies that space on the fine scale is not smooth but full of "bubbles" that wink in and out of existence.

The smallness of the Planck length must give us pause; as much smaller than our present limit as this limit is smaller than us! Plenty of room for more Chinese boxes.

Still, for the first time all our fundamental atoms are pointlike; the proton and neutron were known to have a finite size almost from the outset. So there may be no structure between the quark and the Planck length, though the quarks are far from heavy enough (or impermanent enough) to themselves be part of the quantum foam.

A PHOENIX UNIVERSE?

To close this book, let us keep a promise made at the end of Chap. 12; what is there to learn by slipping quarks into our picture of the big bang?

The first answer is *something about the quarks themselves.* In theory as it stands, the value of α and the masses of the quarks, leptons, and field quanta must be inserted as experimental numbers that cannot be explained. The dream of a field that generates its own matter remains unrealized. One school of thought holds that this can be achieved only by considering *all the fields at once,* and the proper place to do this is in the first instants of the big bang. Then the universe was so small that gravity was as strong as the field that binds quarks.

Perhaps in the beginning there was one great field, encompassing gravity, light, quark binding, and the rest. As the universe expanded, the individual fields "condensed out." Quarks and field quanta, all originally massless, acquired mass, and each field thus acquired its individual character. The subtle interplay of fields in this process has left its mark in the pattern of masses, which can be read as a "recording" of the early moments of the big bang.

A question comes immediately to mind: Did the pattern of masses arise out of logical necessity, or was it simply random?

There are hints in both directions; on the one hand, our present incomplete field theory does not leave the choice of masses *entirely* free; for example, in the weak interaction once the mass of one of the quanta is known, those of the others are completely determined. If a more complete theory finds enough such connections and extends them to quarks, perhaps God really did have no choice after all.

On the other hand John Wheeler insists that during the early moments of the big bang, the universe was so small that quantum fluctations must have played a major role. He would expect the quark masses to be random.

Wheeler's viewpoint takes on a fascinating dimension if it turns out that our universe is fated to stop expanding and fall back in on itself. When it all comes together, it must inevitably reexplode. A new universe is born from the ashes of the old, like the phoenix of legend.

And with the new universe comes *a new toss of the cosmic dice;* The particle masses and with them the structure of matter go up for grabs. The process repeats in endless succession, each new universe as unique as a snowflake, fated to die and be reborn again in fire. In the infinite span of time, *every imaginable universe will get its turn!*

Herein lie the seeds of a possible synthesis of the rival visions of Bohr and Einstein. *God does have choices—an infinity of them—and in the unending fullness of time is obliged to try them all!* There may be reason enough here to satisfy Einstein's yearnings yet chaos enough to delight Bohr's sense of mystery.

How and when (if ever) such outrageous ideas will touch base with experiment is anyone's guess. Each generation of scientists finally comes to the shores of its continent of solid fact. For the time being the ocean beyond can be crossed only in the imagination. This has always been the driving wheel of scientific creativity. The thrill of holding such visions in one's mind is one of the sweetest rewards of the calling of scientist.

An Afterword

To be human is to wonder. Children wonder for a while, before we teach them to be smug about the obvious and to stop asking silly questions. It is easier to pay someone to retain a little of the child and do our wondering for us. We then take comfort in the assumption that anyone devoted to such esoteric pursuits must be insensitive, perhaps even inhuman. With our artists, we perform the equal disservice of regarding them as too sensitive.

Occasionally we are given a glimpse of the finished product. The baby is displayed behind glass, well-scrubbed, and one need not know about the delivery room (it is soundproofed). Thus we are spared the agony of wonder, which is not unlike love and makes as little (or as much) sense as love. But wonder is just too human to fully repress, and it does turn up elsewhere. Some of us turn to fads for the occult, which, interpreted by our twentieth-century minds, becomes a "pop-art" science. More often, we find ourselves left with nothing to wonder about (or to love) but what remains of ourselves after the loss of yet another portion of our humanity.

I, for one, refuse to believe that nothing can be done about this empty place, or about the more general disease of which it is but a minor symptom. But as long as we are sundered so, let me remain one of the children and wonder.

—R.H.M.

Bibliography

CLASSICAL MECHANICS

Dijksterhuis, E. J.: *The Mechanization of the World Picture* (Oxford University Press, London, 1961). A comprehensive history of mechanics up to Newton's time, including ancient and medieval thought.

Galilei, Galileo: *Two New Sciences* (Macmillan, New York, 1914, available in Dover paperback). Galileo's dialogues, elegantly translated by Henry Crew and Alfonso De Salvio.

Jammer, Max: *Concepts of Space* (Harper, New York, 1960) and *Concepts of Force* (Harper, New York, 1962). Two sophisticated monographs on the history of the key concepts of mechanics from ancient to modern times.

Koestler, Arthur: *The Watershed* (Doubleday, Garden City, N.Y., 1960). A heroic biography of Kepler by an author with a somewhat mystical theory of scientific creativity.

Mach, Ernst: *The Science of Mechanics* (Open Court, La Salle, Ill., 1960). A historical and philosophical analysis by the leading champion of positivism in physics.

Newton, Isaac: *Principia* (University of California Press, Berkeley, 1962, 2 vols.). Difficult at points because of the archaic style, but the ideas are presented clearly.

RELATIVITY

Bohm, David: *The Special Theory of Relativity* (Benjamin, New York, 1965). A recent text for physics majors but with considerable emphasis on philosophical aspects.

Eddington, A. S.: *Space, Time, and Gravitation* (Cambridge, New York, 1924). One of the few popularizations of relativity that deals with the general theory.

Einstein, Albert: *Relativity* (Crown, New York, 1931) and *The Meaning of Relativity* (Princeton University Press, Princeton, N.J., 1950). Not the best treatment of relativity for the layman, but refreshingly written and from the horse's mouth. The earlier volume requires less mathematics.

Gardner, Martin: *Relativity for the Millions* (Cardinal, New York, 1965). Perhaps the best popular exposition of relativity.

Schlossberg, Erwin: *Einstein and Beckett* (Links, New York, 1973). Reconstruction of an imaginary, philosophically deep conversation between Einstein and a leading writer roughly his contemporary. Hard reading but rewarding.

Sciama, Dennis: *The Physical Foundations of General Relativity* (Doubleday, Garden, City, N.Y., 1969). A short, readable introduction to the general theory with a stress on its relation to cosmology.

Taylor, E. F., and J. A. Wheeler: *Spacetime Physics* (Freeman, San Francisco, 1963). *The* text on relativity for physics undergraduates. Though some familiarity with the calculus is required, it is beautifully written and treats a large number of examples.

Toben, Bob: *Space-Time and Beyond* (Dutton, New York, 1975). A fanciful extrapolation of the cosmology of John Wheeler that attempts to build a scientific base for ESP and other occult phenomena.

QUANTUM THEORY

Bohm, David: *Causality and Chance in Modern Physics* (Routledge, London, 1957). A critique of the Copenhagen interpretation of the quantum theory by its most persistent contemporary critic.

Bohr, Niels: *Atomic Physics and Human Knowledge* (Vintage, New York, 1966). Bohr's views on the broader implications of his physical thinking.

Broglie, Louis de: *The Revolution in Physics* (Noonday, New York, 1953) and *The Current Interpretation of Wave Mechanics: A Critical Study* (Elsevier, Amsterdam, 1964). Popular expositions of modern physics, with the author's views as to its philosophical implications.

Cline, Barbara: *Men Who Made a New Physics* (Signet, New York, 1966). Biographies of all the leading figures in the development of the quantum theory, with some exposition of their ideas.

Forman, Paul: Weimar Culture, Causality, and Quantum Theory 1918–1927: Adaptations by German Physicists to a Hostile Intellectual Environment, in *Historical Studies in the Physical Sciences*, vol. III (University of Pennsylvania Press, 1971). The title is self-explanatory.

Jammer, Max: *The Conceptual Development of Quantum Mechanics* (McGraw-Hill, New York, 1966). The definitive history of the quantum theory, by an author who is both an able historian and a theoretical physicist. Extremely thorough, difficult for students weak in mathematics, but worth the effort.

Moore, Ruth: *Niels Bohr* (Knopf, New York, 1966). A definitive biography of Bohr, with an excellent exposition of the early quantum theory.

Toulmin, Stephen, and June Goodfield: *The Architecture of Matter* (Harper, New York, 1962). An excellent history of atomism and other ideas related to the structure of matter, from ancient to modern times.

OF GENERAL INTEREST

Beiser, Arthur (ed.): *The World of Physics* (McGraw-Hill, New York, 1960). Selections from popular or semipopular writings of noted physicists, with the emphasis on the modern era.

Čapek, Milič: *The Philosophical Impact of Contemporary Physics* (Van Nostrand, New York, 1961). Impact of relativity and quantum theory on traditional philosophy, from an historical perspective.

Holton, Gerald: *Thematic Origins of Scientific Thought, Kepler to Einstein* (Harvard University Press, Cambridge, Mass., 1973). Thoughtful, scholarly, but readable essays on Kepler, Bohr, and Einstein, with an emphasis on the cultural and psychological roots of their ideas.

Kuhn, Thomas S.: *The Structure of Scientific Revolutions, 2d ed.* (University of Chicago Press, Chicago, 1970). A modern treatise on philosophy of science that emphasizes historical factors.

Shamos, Henry: *Great Experiments in Physics* (Holt, New York, 1960). Descriptions of some of the key experiments in the history of physics, from Galileo to the present.

Watson, James D.: *The Double Helix* (Signet, New York, 1968). An exciting, personal, and controversial account of what it means to participate in an epoch-making scientific discovery.

Appendix

Chapter 1

INTERPRETING A TABLE OF DISTANCE VERSUS TIME

	Measured	Computed	
Time, s	Distance, m	Speed, m/s	Acceleration, m/s^2
0	0	5	
2	10	7	1
4	24	9	1
6	42	11	1
8	64		

The table above is an example of one way to study motion. The distance traveled is recorded at equal time intervals.

The average speed is easy to compute for each interval. For example, between 2 and 4 s, a time interval $\Delta t = 4 - 2 = 2$ s, and the body traveled $\Delta s = 24 - 10 = 14$ m. By the definition of average speed,

$$\bar{v} = \frac{\Delta s}{\Delta t} = \frac{14}{2} = 7 \text{ m/s}$$

This operation shows the significance of the Δ notation. It means that the values used in the formula were obtained by subtracting the value at the beginning of the interval from that at the end.

The reader is invited to verify that the other entries in the speed column are correct

Moving on to the acceleration, note that the speed increases from 5 to 7 m/s in the time between the 0- to 2-s interval and the 2-to 4-s interval. Since this represents a change of 2 m/s in a 2-s interval, the acceleration is

$$\bar{a} = \frac{\Delta v}{\Delta t} = 1 \text{ m/s}^2$$

Since the speed increases by 2 m/s each time, the acceleration is uniform. In this case, it is easy to relate these average values of speed and acceleration to their instantaneous values. Since the body gains an equal amount of speed in each 2 s of the interval, the average value must represent the speed at the midpoint of the interval; i.e., the instantaneous speeds, 5, 7, 9, 11 m/s were reached at $t = 1, 3, 5, 7$ s. But the acceleration may not have been constant—indeed the calculation of the acceleration would have been only approximate, since we would not know the exact time interval required for a change in speed.

Of course, we are taking on faith that the acceleration stayed the same throughout each interval.

QUESTIONS

1. Cite an example of two objects where the heavier actually falls more slowly than the lighter.

2. As an additional argument against Aristotle, Galileo cited the example of the fall of two identical objects attached by means of a string. Try to concoct such an argument.

3. Invent an example (in addition to that cited in the text) of a common experience which indicates that falling bodies do not acquire speed instantaneously.

4. Give an argument to show that if an object moves with *increasing* (rather than uniform) acceleration, its average speed is *less than* half its final speed.

EXERCISES

Note: In this and all subsequent sets of exercises, use the approximate value for the acceleration due to gravity

$$g = 10 \text{ m/s}^2$$

rather than the more precise 9.8 m/s², except where specifically directed to use the more precise value. This value saves on arithmetic. In fact, because of differences in altitude, the rotation of the earth, and other local effects, g varies from place to place nearly as much as the difference between the "precise" and approximate values.

1. A fast ocean liner makes the crossing from New York to Southampton, a distance of 3400 mi, in 4 days 4 h. What is the average speed of the liner?

2. A fast sprinter can cover 100 m in 10 s flat.
    (a) What is the average speed of the sprinter?
    (b) What would be his time for the mile (1610 m) if he could keep up a sprint pace?

3. A sky diver bails out at an altitude of 3000 m (about 10,000 ft). After the first few seconds of the jump, he drops at a constant speed of 40 m/s until he reaches a height of 200 m, where he opens his parachute, which drops his speed of fall to 10 m/s.
    (a) Calculate how long it takes him to reach the ground.
    (b) Calculate his average speed for the entire time of fall.
    (c) Comment on whether the calculation in part (b) is particularly meaningful.

4. A sports car can accelerate at 5 m/s². How long does it take to go from 0 to 30 m/s (about 70 mi/h)?

5. A car speeds up from 24 to 30 m/s in 2 s. Calculate its acceleration.

6. An object initially moving 20 m/s experiences a negative acceleration of 2 m/s² for 3 s. What is its final speed?

7. How long does it take for a heavy object dropped from the top of the Empire State Building (320 m high) to reach the ground?

8. Galileo's inclined plane was about 4 m long, and it took a ball about 10 s to roll the full length of the plane. Calculate the acceleration and compare it with that of a freely falling body.

Chapter 2

A PRACTICAL FORMULA FOR PROJECTILE RANGE

The discussion in Chap. 2, while revealing the physics behind projectile motion, is not of much practical value because one does not ordinarily separately measure the horizontal and vertical components of a projectile's velocity. Ordinarily, one knows the *speed* and the *starting angle*. For example, a gun always fires a bullet at nearly the same speed. To use these variables, we must make use of a trigonometric relationship, illustrated in the diagram.

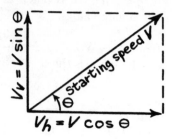

θ is referred to as the *angle of elevation*. Then the equation becomes

$$R = 2\frac{V_v V_h}{g} = \frac{2(V \cos \theta)(\sin \theta)}{g}$$

$$= \frac{2V^2 \cos \theta \sin \theta}{g}$$

Those with training in trigonometry may remember the identity

$$2 \cos \theta \sin \theta = \sin 2\theta$$

Thus the formula becomes

$$R = \frac{V^2 \sin 2\theta}{g}$$

From inspection of this formula, two facts about projectiles become apparent:

1. Since the sine of an angle achieves its maximum value (1.0) at 90°, sin 2θ is a maximum at 45°. Thus, a projectile achieves its greatest range for an angle of elevation of 45°.

2. From the fact that sin $x = \sin(180° - x)$ we can see that sin $2\theta = \sin(180° - 2\theta) = \sin[2(90° - \theta)]$. Thus, a projectile with an angle of elevation of 10° travels the same distance as one with elevation 80°. Of course, the trajectories look very different in the two cases.

QUESTIONS

1. Discuss the assertion that "the more education people have, the faster they learn" as evidence for the existence (or nonexistence)

of a principle in psychology of learning analogous to the principle of superposition.

2. Propose a law analogous to a conservation law for a field outside the natural sciences.

3. A ball rolls down a wide inclined plane. Instead of starting from rest, it is given an initial *horizontal* push (perpendicular to the line down which it would otherwise roll). What path would the ball trace on the plane? Will the time required to reach the bottom be affected?

4. Cite at least three common everyday experiences that illustrate the operation of momentum conservation.

EXERCISES

1. A man on a moving train drops a ball (inside the train). A stationary observer on the ground observes the process through the train window.
   (a) Sketch the path of the ball, as seen by each observer.
   (b) Repeat part (a) for the case where the train is *accelerating*.

2. A car drives off a sheer cliff, moving horizontally at 30 m/s. It strikes the ground 2 s later.
   (a) How far from the base of the cliff does the car land?
   (b) How high is the cliff?

3. A projectile travels a distance of 200 m in 4 s.
   (a) What was its horizontal velocity component?
   (b) What was its vertical velocity component?
   (c) How high was it at the top of its trajectory?

4. A clay ball of mass 3 kg and velocity 16 m/s strikes a stationary ball of mass 5 kg and sticks to it. What velocity does the combined mass have, after the collision?

5. Two clay balls of equal mass are moving in opposite directions when they collide and stick together. Before the collision one ball is moving 7 m/s to the right and the other is moving 3 m/s to the left. What is the velocity and direction of the combined mass after the collision?

6. A ball of mass 2 kg and velocity 10 m/s strikes a stationary ball of mass 1 kg. After the collision, the 2-kg ball is moving at 4 m/s in its original direction.
   (a) What are the speed and direction of the 1-kg ball after the collision?
   (b) Is this an example of an elastic collision?

7. A sealed freight car 40 ft long and weighing 4000 lb when empty

sits on a frictionless track. The car starts to move and finally stops after moving 1 ft.

(a)   Is this sufficient information to conclude that something is moving around inside the car?

(b)   If the motion is due to something moving inside the car, give numerical proof that the motion of one person walking around in the car could account for the motion observed.

Chapter 3

QUESTIONS

1.   The normal method of comparing masses to a standard is to compare weights on a *balance*. What assumptions must be made for this to be a valid way of comparing masses?

2.   A man swings a bucket of water in a vertical circle. If the motion is sufficiently fast, the water does not spill at the top when the bucket is upside down. Explain why this is possible.

3.   Automobile racers ordinarily try to take a curve by starting at the outside lane, cutting to the inside, and ending up again at the outside. Explain why they do this.

EXERCISES

1.   If a 3-kg mass is observed to be speeding up at a rate of 5 m/s², how large a force (in newtons) must be acting on it?

2.   (a)   Convert your own mass into kilograms (1 kg = 2.2 lb).
     (b)   Calculate the force in newtons exerted by gravity on your body.

3.   A car is being towed by a rope which will break if it transmits a force of more than 5000 N. The mass of the car is 1000 kg. What is the maximum acceleration that can be given to the car in this way?

4.   A car which can normally accelerate at 4 m/s² is towing an identical car. What is the greatest acceleration it can achieve?

5.   Show that a jet airliner which flies twice as fast as a propeller plane will turn in a circle four times as big and take twice as long to complete a turn through a given angle as the propeller plane at the same acceleration.

Chapter 4

THE MATHEMATICAL RELATION BETWEEN KEPLER'S THIRD LAW AND INVERSE-SQUARE FORCES

For circular orbits, Kepler's third law relates the length of the year T to the radius R of the orbit:

$$T^2 = \text{const} \times R^3$$

To see the connection with the force, we must see what this law says about the *acceleration* of the planet:

$$a = \frac{V^2}{R}$$

Since V is the circumference of the orbit, $2\pi R$, divided by the length of the year, we have

$$a = \frac{(2\pi R)^2}{T^2} \frac{1}{R}$$

But $T^2 = \text{const} \times R^3$. Lumping this constant and the factor $4\pi^2$ into a new constant, we have

$$a = \text{const} \times \frac{R^2}{R^3 \times R} = \frac{\text{const}}{R^2}$$

Thus, the acceleration is proportional to the inverse square of the distance. For elliptical orbits the situation is more complex, but Newton was able to show that the same relation holds. (In this case, Kepler's third law replaces the radius by one-half the length of the *major axis*, the long dimension of the ellipse.)

QUESTIONS

1. One additional support for Newton's law of universal gravitation was that, in his time, the four largest moons of the planet Jupiter had been discovered and their motion obeyed Kepler's laws. Comment on what support this lends to Newton's theory.

2. Although the gravitational constant was still unknown in Newton's day, and thus the mass of the earth was unknown, one could still make a comparison between the mass of the earth and that of the sun. Explain how this was possible.

EXERCISES

Planet	Distance from sun, earth orbit radii	Length of year, earth years
Mercury	0.39	0.24
Venus	0.72	0.61
Earth	1.00	1.00
Mars	1.52	1.9
Jupiter	5.2	12

1. Verify that the planetary data in the table above satisfy Kepler's third law.

2. (a) Calculate the acceleration of the earth in its orbit around the sun, using the fact that its orbital velocity is 30,000 m/s and the orbit radius is 1.5×10^{11} m. (If you are unfamiliar with scientific notation, see footnote on page 70.)

 (b) Compare the result to the acceleration due to the earth's own gravity at its surface.

 (c) Use the result to calculate the ratio of the sun's mass to that of the earth.

Chapter 5

QUESTIONS

1. A cannonball is fired in a high trajectory, is slowed by air resistance, and eventually buries itself in the ground. Describe the series of energy changes that take place from the moment the powder is ignited.

2. Invent a conservation law (it need not actually be true) analogous to energy conservation in a field outside the natural sciences.

3. A pendulum consisting of a lead weight on a light flexible string is released from the point shown in the drawing. At the bottom of the pendulum's swing, the string strikes a rigid, immobile peg placed as shown. Assume energy is conserved.

 (a) Copy the sketch and show where the ball will rise to at the top of its swing to the right.

 (b) Show on the same sketch where the peg should be placed if you want the string to wind around the peg.

(c)   What statement *about the peg*, made above, assured you that it would not interfere with energy conservation? Explain.

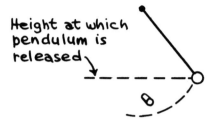

Height at which
pendulum is
released

4.   Contrast the physical definitions of the words *work, power, force,* and *energy* with their ordinary-language definitions. Do you think the choice of terms was apt or farfetched?

EXERCISES

1.   How many joules of work must be done to lift a 5-kg object a height of 10 m?

2.   A force of 5 N pushes on an initially stationary object of mass 4 kg for a distance of 10 m.
     (a)   Calculate the work done.
     (b)   Calculate the final speed of the object.

3.   An elastic collision, viewed in terms of the energy concept, proves to be one in which both kinetic energy and momentum are conserved. Refer back to the example of an elastic collision in Chap. 2 and verify that kinetic energy is conserved.

4.   A car of mass 1000 kg speeds up from 20 to 30 m/s in a distance of 200 m.
     (a)   How large a force does this take?
     (b)   What is the acceleration of the car while it is speeding up?

5.   A bicyclist approaches a hill, moving 10 m/s. How far vertically can he coast without pedaling?

6.   It takes 4120 J of energy to heat 1 kg of water by 1°C. If all the energy is converted to heat in the water, how large a rise in temperature occurs in water going over a 100-m waterfall?

7.   A 2000-kg car is capable of maintaining a speed of 15 m/s while climbing a street which rises 1 m for every 5 m of length. How many watts of power are required? Convert the result to horsepower.

Chapter 6

QUESTIONS

1. The field concept is most useful for explaining the forces on small charges in the presence of larger ones. Explain why.

2. Gravity, like electricity, can be interpreted as a field. Write the formula for the gravitational field strength. To what physical quantity, mentioned in this text, is gravitational field strength identical?

3. Today we also believe gravity to propagate at the speed of light. In view of this fact, why was it possible for Newton to overlook this feature and still explain planetary motion?

4. One way of resolving the "nightmare of determinism" is to prove that a calculating machine capable of analyzing the entire universe in this detail would have to be as complex as the universe itself. Construct an argument to show this is so. State whether, in your opinion, this resolves the problem.

Chapter 7

QUESTIONS

1. Devise instructions for the marching band used as a wave example in Chap. 8 that will enable one to use them to demonstrate the principle of superposition.

2. A guitarist can obtain a very pure tone one actave above the fundamental tone of a string by touching it very lightly at its midpoint. Explain why this tone should be pure (not mixed with other wavelengths) and why he must use the fingerboard to obtain most other notes.

3. A radio station is required to design an antenna that will avoid transmitting to the south, where another station of the same frequency is located. The engineer designing the station builds an antenna consisting of two towers $\frac{1}{2}$ wavelength apart, separated along a north-south line. Explain.

4. A satisfactory (but rather slow) camera can be made with a pinhole instead of a lens. As the pinhole is made smaller, at first it tends to make the image sharper. Eventually, however, too small a pinhole will produce a blurred image. Explain.

EXERCISES

1. If a wave of frequency 10 Hz has a wavelength of 100 m, what is the wave velocity?
2. The velocity of light is 3×10^8 m/s, and the frequency of a typical FM radio station is 10^8 Hz. What is the wavelength of the waves broadcast by this station?
3. What are the wavelengths of the three longest standing waves that can exist on a string of length 3 m?
4. Two hi-fi speakers placed 3 m apart are sounding a note of wavelength 2 m.
   (a) Show that if one walks along a line 4 m in front of the speakers, there will be near silence at the points directly in front of them.
   (b) In all, at how many points along the line will you hear a maximum of sound intensity?

Chapter 8

A USEFUL APPROXIMATION

When using the formulas derived in Chap. 8, one frequently deals with situations where v is much less than c. In these cases it is handy to know the following approximate formulas:

$$\sqrt{1 \pm x} \approx 1 \pm \tfrac{1}{2}x \qquad \frac{1}{1-x} \approx 1 + x$$

The wiggly equals sign means that the formula is not exact.

The first approximation is surprisingly good even when x is not much less than 1. For example, we have, using this formula,

$$\sqrt{2} = \sqrt{1+1} \approx 1\tfrac{1}{2}$$

which is within 6 percent of the correct answer, 1.414. By the time we get to

$$\sqrt{1.21} = \sqrt{1+0.21} \approx 1.105$$

the answer is very close to the true one, 1.100.

Similarly, for the second formula, we have

$$\frac{1}{0.8} = \frac{1}{1-0.2} \approx 1 + 0.2 = 1.2$$

which is not far from the correct answer of 1.25.

Combining the two approximations, we obtain

$$\gamma = \frac{1}{\sqrt{1 - (v/c)^2}} \approx \frac{1}{1 - \frac{1}{2}(v/c)^2} \approx 1 + \tfrac{1}{2}(v/x)^2$$

For the Michelson-Morley case, where the expected value of v/c was around 10^{-4}, the formula can be regarded as nearly exact.

EXERCISES

1. A ferryboat, which can cruise at 13 mi/h, crosses a river 2.4 mi wide, in which a current of 5 mi/h is flowing. Calculate the time for the trip.

2. From the approximate formula given at the beginning of this section, calculate values of γ for $v/c = 0.1, 0.2, 0.3, \ldots, 0.8$. Make a graph of the results, plotting on the same graph the result from exact arithmetic (Fig. 8-2).

3. Suppose the Michelson-Morley experiment is performed with light of wavelength 4×10^{-7} m on an interferometer that gives a round-trip path length of 12 m in each arm (these figures are close to those for the actual instrument built in Cleveland). How many wavelengths' shift would be expected if the aether wind were 30 km/s $= 10^{-4} c$?

4. Using the result of Exercise 3, what is the smallest aether-wind speed that can be observed if it is possible to detect a fringe shift of one-twentieth of a fringe?

Chapters 9 and 10

RELATIVISTIC VELOCITY ADDITION

At the beginning of Chap. 9, the paradox of Einstein's postulate was phrased as follows: "To the man on the train, light behaves like a bullet, while to the man on the ground it behaves like the sound of the shot." The resolution of this paradox was suggested at the end of that chapter, in which it was pointed out that, from the point of view of the man on the ground, the man on the train is using clocks that are slow and unsynchronized to measure motion over a contracted length, and thus gets an exaggerated figure for the speed of the "bullet" with respect to *him*.

Stated quantitatively, relativity says it is not correct to simply add the speed of the bullet relative to the train to the speed of the train relative to the ground to obtain the speed of the bullet relative to the ground. If v is the speed of the train and u the speed of the bullet measured by the moving observer, the observer on the ground finds the bullet moving at the speed

$$\frac{u + v}{1 + uv/c^2}$$

where our commonsense notion would be that the answer is $u + v$.

A little playing with this formula will reveal its significance:

1. If both u and v are small compared with c, the term uv/c^2 is much smaller than 1. Thus, the formula reduces to the commonsense one for ordinary velocities, for the denominator is essentially 1.

2. If *either* u or v is equal to c, the formula gives the answer c, as a small amount of algebra will show:

$$\frac{c + v}{1 + cv/c^2} = \frac{c(1 + v/c)}{1 + v/c} = c$$

This resolves the paradox: By the relativistic rule, c *plus anything is c.* Thus, both the stationary observer and the moving one get the same value for the speed of the bullet, if it is moving at the speed of light.

QUESTIONS

1. A spaceship passes a "stationary" observer. Both he and the spaceship crew are provided with identical timers, which they start at the instant they pass. Each timer is wired to a light, which it flashes after 10 s has elapsed. Which light does the stationary observer believe flashes first? Which one does the spaceship crew believe flashes first? Give a nonmathematical argument to explain why their points of view are not inconsistent.

2. The discrepancies between what two observers relatively in motion see arise not from what they actually see but from their interpretations of their observations, each in his own frame of reference. Invent an example of such a situation outside the physical sciences.

EXERCISES

1. A 1-m rod moving at $0.8c$ would appear how long to a stationary observer?

2. How much is an actual train 1000 m long moving at 30 m/s shortened? Use the approximate formula given in the appendix to Chap. 8.

3. How fast would a train have to be moving in order to shorten its length by 2 percent? (Use the same approximate formula.)

4. How much slower does a clock moving at 0.6c run than an identical clock standing still?

5. A train 1000 m long is moving at 0.9c. What is the difference in clock readings at the two ends of the train, according to an observer on the ground?

Chapter 11

KINETIC ENERGY

The approximate formula stated in the Chap. 8 section of the Appendix can be used to show that the definition of kinetic energy as $\frac{1}{2}mv^2$ still works if v is much less than c. In this case we have

$$E = mc^2 = \gamma m_0 c^2 = \left(1 + \frac{1}{2}\frac{v^2}{c^2}\right)m_0 c^2$$
$$= m_0 c^2 + \frac{1}{2}m_0 v^2$$

where m_0 is the *rest mass*.

The term $m_0 c^2$ represents the *rest energy* of the object, which it possesses just by existing. The term $\frac{1}{2}m_0 v^2$ represents the additional energy due to motion, ie., the kinetic energy.

QUESTIONS

1. State whether there would be a change in mass and which way the change would go in each of the following situations:
   (a) An automobile battery is charged.
   (b) A red-hot steel bar is allowed to cool down.
   (c) A rubber band is stretched.
   (d) Hydrogen and oxygen burn to form water inside a sealed, insulated container.
   (e) Same as part (d) in a container that allows the heat to escape.
   (f) Two atoms joined together by a force of attraction are pulled apart.

EXERCISES

1. Using the approximate formula, calculate the increase in mass of a jet airliner of mass 100,000 kg traveling at its normal cruising speed of 300 m/s. Is this a measurable effect?

2. How fast must an object be moving in order to have a mass double what it has standing still?

3. Calculate how many joules of energy would be produced by converting 1 kg of mass to some other form of energy.

4. In a nuclear reaction, about one-thousandth of the mass of the fuel consumed is converted into energy. A nuclear power plant is designed to provide 10^{11} W of electricity (enough for a city of about 100,000). Assume the power plant delivers as electricity one-third of the energy it produces. How much fuel does it consume in a year of operation (1 year is about 3×10^7 s)?

Chapter 12

QUESTIONS

1. General relativity reduces the old argument between an earth-centered and a sun-centered picture of the solar system to a nearly arbitrary choice. Explain why.

2. With a world globe and a string, verify that the shortest route from New York to Rome does not follow the parallel (due east-west line) on which they both lie. How far north does the route go?

EXERCISES

1. Let the astronaut in the twin paradox send a message home half-way through his outbound voyage.
   (a) Adopting the earth-bound viewpoint, when is the message received?
   (b) How does the astronaut account for the result of part (a)?

2. How long does it take light to travel 3 km? How far would a falling body drop in this time?

3. Calculate the difference in clock rates for two clocks separated vertically by 20 m at the earth's surface. (Small as the answer is, it can be and has been measured!)

Chapter 13

DETAILS OF THOMSON'S EXPERIMENT

The forces on charged particles are easy to calculate, once the strength of the electric and magnetic fields is known. In an electric field,

$$F = eE$$

where e is the charge of the particle and E the electric field strength. The force of a magnetic field of strength B on a particle moving with velocity v perpendicular to the field is

$$F = evB$$

Thomson rather cleverly arranged to simultaneously apply electric and magnetic fields that deflected the cathode rays *in opposite directions*. The fields were adjusted until there was no deflection of the beam. Thus, he knew the electric and magnetic forces were exactly equal:

$$eE = evB$$

with the result that, regardless of the size of e, he knew the velocity of the particles:

$$v = \frac{E}{B}$$

Knowing the velocity, he knew how long the particles spent in the field. He then turned off the magnetic field, measuring the deflection of the beam in the electric field alone. The mathematics of this deflection were precisely those of *projectile motion*—the electric force is perpendicular to the original line of flight. From the observed deflection, Thomson could calculate the acceleration due to the electric force. Knowing this, he could find the ratio of charge to mass:

$$F = ma = eE$$

$$\frac{e}{m} = \frac{a}{E}$$

QUESTIONS

1. Invent explanations or descriptions in terms of the atomic theory of matter for the following phenomena:

(a) The boiling of a liquid

(b) Surface tension in liquids

(c) The fact that when a gas is compressed by a piston, it gets hotter

2. If a small drop of oil is placed on a large, smooth surface of water, the oil film will not spread out indefinitely but will spread out to form a circle, the area of which is proportional to the amount of liquid in the drop. Give an argument using this fact as evidence for the atomic theory of matter and devise a procedure whereby this phenomenon can be used to estimate the size of oil molecules.

3. There is no known substance that remains a gas at absolute zero, and all gases deviate from the simple behavior of Fig. 13-2 when they are cooled close to the temperature at which they condense into a liquid. Do these facts in any way undermine the atomic theory of matter? Explain.

4. Amplify the argument by which Faraday's law of electrical conduction in liquids was used to conclude that all ions carry the same electric charge.

5. Cite the evidence that the electron is a very light particle carrying the same charge as an ion rather than an object as heavy as an atom carrying a much larger charge.

6. State in your own words what you consider the most convincing evidence for the atomic nature of matter.

EXERCISES

1. Using the data on atomic weights given early in Chap. 13, give the "recipe" (ratio of weights) for forming hydrochloric acid (HCl) from hydrogen and chlorine.

2. Room temperature is normally about 20°C. What is this temperature on the Kelvin scale, which is Celsius degrees above absolute zero?

3. If a closed container holds gas at 0°C, to what temperature must it be heated to double its pressure?

4. Making use of the fact that gas temperature is proportional to the average kinetic energy of the molecules and the proportionality is the same for all gases, find the ratio of *speeds* of hydrogen and oxygen molecules, at the same temperature. (Each gas forms molecules consisting of two atoms.)

5. Refer back to Question 2 above. Suppose a droplet of oil of volume 0.001 cubic centimeters (cm^3) spreads out on the surface of a large tank of water to form a circle 40 cm in radius. Estimate the size of the oil molecules.

Chapter 14

QUESTIONS

1. From the behavior of atoms in liquids and solids, it is clear they are pretty much "hard spheres"; i.e., once they actually come in contact, there are strong forces of repulsion that tend to keep them from penetrating one another very far. Discuss the problem of explaining this for the plum-pudding atom, in which the positive sphere can be freely penetrated by electrons, and the planetary atom, which is mostly empty space.

2. There are a number of light sources in ordinary use that produce light in a low-density gas; neon signs and the bluish mercury-vapor lamps used for street lighting are examples. Observe such lights through a glass prism. How many "lines" do you see?

3. Compare Rutherford's use of the Geiger-Marsden data to justify his nuclear atom with Newton's use of Kepler's laws to justify the inverse-square law of gravity.

EXERCISES

1. The diameter of a gold atom is about 2.5×10^{-10} m. Rutherford's gold foil was about 10^{-6} m thick. About how many gold atoms did each alpha particle that penetrated the foil traverse?

2. In order to come off back of 90°, an alpha particle would have to pass within 10^{-13} m of the nucleus. Imagine each nucleus to be surrounded by a circular "target" of this size. If the alpha comes that close, it will come off back of 90°.
   (a) In an encounter with a single atom of gold, what are the chances for the alpha to come off back of 90°?
   (b) Using the result of Exercise 1, about what fraction of the alphas that struck the foil came off back of 90°?

3. Construct an ordinary (nonlogarithmic) graph of the Geiger-Marsden data. This should convince you of the value of the logarithmic graph.

Chapter 15

QUESTIONS

1. It is far easier to demonstrate that electromagnetic radiation is quantized if x-rays or gamma rays are used than with visible light. Why?

2. One later confirmation of Einstein's analysis of the photoelectric effect involved the following experiment. The photoelectric effect is produced by using a very faint light and a rapid camera shutter. The light is so faint that the total energy that should pass through the shutter is less than that required to pull an electron free from the metal. It is found that, if the shutter is clicked many times, occasionally an electron is ejected from the metal. What would the classical picture of light predict? Show how the photon theory can account for the phenomenon.

3. Which two adjoining Bohr orbits are farthest apart in energy? Explain.

4. Describe how collisions between atoms make it possible for a hot gas to give off light, in terms of the Bohr picture of the atom.

5. List all the physical laws used in the Bohr picture of hydrogen, classifying them as
   (a) Carried over from Newton's mechanics
   (b) Originating from earlier quantum theory
   (c) Bohr's own contribution

6. If Bohr's model is applied to atoms heavier than hydrogen, with greater positive charge on the nucleus, would you expect the innermost orbit to be smaller or larger than that in hydrogen? Explain.

EXERCISES

1. What is the ratio between the energy of a photon given off when an electron falls from the third to the second Bohr orbit, to that given off when it falls from the second to the first?

2. In a photoelectric effect experiment, the following results are obtained: the lowest frequency that can produce electrons is 10^{15} Hz; and with light of 3×10^{15} Hz, the electrons have an energy of 12.8×10^{-19} J. Use these data to obtain (a) Planck's constant and (b) the energy required to free an electron from the metal.

Chapter 16

QUESTIONS

1. As one moves to larger and larger Bohr orbits, do the wavelengths of the electrons get larger or smaller?

2. If a photon and an electron have the same kinetic energy, which has the shorter wavelength?

3. The de Broglie formula implies that the heavier a particle is, the shorter its wavelength. Does this make it easier or harder to account for the absence of wave effects with particles of visible size?

4. Which undesirable features of the Bohr atomic model are eliminated in the wave model? Which remain? What new problems does the wave model introduce?

5. Comment on whether the success of the de Broglie theory makes Planck's constant seem more or less universal in its significance than before the theory was suggested.

6. In the Bohr theory, Planck's constant plays two distinct roles: in one it connects the energy states with the frequencies of spectral lines, while in the other it determines the allowed orbits. Explain how the de Broglie theory unifies these roles.

EXERCISES

1. What is the ratio between the wavelengths of de Broglie waves of electrons in the second and first Bohr orbits? (*Hint:* Do not actually calculate the momentum; just consider the size and wave pattern of each orbit.)

2. Calculate the wavelength of a running man of mass 100 kg at speed of 3 m/s. Is the result reassuring or disturbing?

3. Show that the de Broglie formula gives the correct wavelengths for photons, using the fact that the energy carried by light is its momentum times the speed of light.

4. Electrons (mass 9×10^{-31} kg) in a TV picture tube travel at $0.2c$. Using nonrelativistic formulas:
   (a) Calculate their momentum, their kinetic energy, and their wavelength.
   (b) Compare the wavelength to that of a photon of the same energy.

Chapter 17

QUESTIONS

1. Explain why the more massive an object is, the easier it becomes to predict its future position.

2. The velocity of an electron is measured by timing its flight between two viewing screens whose positions are well known; that is, Δx is small and Δp is large. Discuss what the uncertainty principle has to say about our knowledge of

   (a) The velocity of the electron during its passage between the two screens

   (b) The velocity of the electron after leaving the second screen

   (c) The velocity of the electron before striking the first screen

   (d) The agreement among many measurements of part (a) if the electrons originate from a source that gives them all the same momentum before striking the first screen.

EXERCISES

1. Calculate the uncertainty in momentum of an object whose position is measured to an accuracy comparable to the size of an atomic nucleus (10^{-14} m).

2. Verify the entries in the last two columns of Table 17-1.

Chapter 18

QUESTIONS

1. Criticize the following naïve statement of the wave-particle duality: "The true nature of an electron is intermediate between that of a particle and of a wave."

2. The quantum theory contains a wave-particle duality, and relativity has a space-time duality. In both cases, it is the *observer* that influences which aspect manifests itself. Discuss the respects in which these two dualities are similar and different.

Chapter 19

QUESTIONS

1. Find a suitable reference, e.g., an encyclopedia yearbook, a *Scientific American* article, or one suggested by your instructor, and report on progress in the quark theory since 1976.

2. Which of the two possibilities for the universe, logical necessity or random but infinitely repeated form, do you find most appealing? Or have you an alternative you prefer?

EXERCISES

1. An electron and positron annihilate, producing a temporary photon that materializes a mu-plus–mu-minus pair. In this case, no inter-action with a nucleus is necessary.
   (*a*) Draw the Feynman diagram for the process.
   (*b*) Explain why no nucleus is necessary. (*Hint:* The photon is in existence for a very short time.)

2. Some theorists expect that in its final version, the quark theory will have six rather than four flavors for quarks and leptons. How many mesons and baryons would then be possible?

Index

Index

E

S

T

U

W

Z